The Fire Service:
History, Traditions
&
Beyond

By

J.A. Rhodes

Booklocker.com, Inc.
2006

Acknowledgements

This book would not have become reality if it were not for the hard work and unselfish dedication of David Lewis. On top of an already full plate, he guided me through every step of the way by reviewing the book for historical accuracy, gathering pictures, and answering all questions. Everyone should be grateful for the passion he has towards the fire service and the Aurora Fire Department. He is without question a gift that should not be taken for granted.

A special thanks goes out to Annie McCue for editing this book. At a time when I thought that I would not find someone to help me with this process, she came through for me. Thank you.

Dedication

This book is dedicated to all those who have served before us without due regard for their own life, and the love they had, for the lives they chose to serve.

Disclaimer

Do you know the funny thing about history? It's history. When you go to sleep tonight and wake up in the morning, the day before has now become history. The events that occurred just a mere twenty-four hours earlier can be misconstrued and the proverbial history ball starts to roll. The path that this so-called history ball takes is a delicate balance on a knives' edge somewhere between truth (what actually happened) and perception of reality (what everyone thinks happened). Remember; when it comes to stories there are three different versions, your side, their side and the ultimate truth.

The problem with this is that when a significant event occurs there will always be some variation in exactly what took place. People throughout history had many claims that they were the "first" to discover something. Just as Christopher Columbus thought he had discovered the new world only to find out that the Native Americans were really there "first". Somebody somewhere could have been the "first" to do something, whatever it may be, but if there was nobody there to document the event then it could have gone unnoticed. Folklores, stories, and legends all have contributed to fogging historical events in one way, shape, or form. What we do know and believe is what we read in textbooks, documents, and so on. The information that you will read is as accurate as what history allows it to be.

Forward

The fire service is rich in its history and tradition, but often goes unrecognized and taken for granted. The inspiration of this book was fueled by my own embarrassment when I realized that I did not know the history of the fallen firefighters on my department. The shame soon encompassed the simple traditions of the fire department that I should have known, but did not. This is when I realized that if I wasn't aware, then most likely others where oblivious as well. Something had to be done to remedy the problem.

The project started off easy enough, but a trip to Boston would turn the simple project into what you are holding now. Standing kitty-corner to where the Vendome Hotel once stood, I read the story of the nine firefighters that lost their lives on that June day. These men gave their lives protecting the citizens of Boston and are now honored by a memorial that is silently saying we will never forget you.

There have been two dark days for the Aurora Fire Department, first on May 9th, 1929 and again on January 11th, 1934 where firefighters lost their lives. No memorials where erected to pay homage to these fine men. All proceeds from this book are going to the purchase of plaques that will honor the fallen firefighters of Aurora and to the building of the new fallen firefighter's memorial that will help us remember all that made the supreme sacrifice. This is the history and traditions of the fire service.

TABLE OF CONTENTS

I HAVE NO AMBITION BUT ONE

I HAVE NO AMBITION IN
THIS WORLD BUT ONE, AND
THAT IS TO BE A FIREMAN. THE POSITION MAY, IN
THE EYES OF SOME, APPEAR
TO BE A LOWLY ONE; BUT THOSE WHO KNOW THE
WORK WHICH A FIREMAN
HAS TO DO BELIEVE
HIS IS A NOBLE CALLING.
OUR PROUDEST MOMENT
IS TO SAVE... LIVES.
UNDER THE IMPULSE
OF SUCH THOUGHTS
THE NOBILITY OF THE
OCCUPATION THRILLS US
AND STIMULATES US TO
DEEDS OF DARING,
EVEN OF
SUPREME SACRIFICE.

Edward F. Crocker
Chief of Department
FDNY
1899-1911

Introduction
IN THE BEGINNING

What do you want to be when you grow up? I want to be a fireman. This childhood dream of riding on the big red trucks with the wailing sirens and flashing lights has become a reality for some but yet still a vision for others. There is a saying, "If you love what you do, you'll never work a day in your life". Now who said this to be exact I have no idea, but there isn't a truer statement ever.

What draws one to such a profession? Is it the dream of one day becoming a "hero" just like they have seen on TV or in the movies? Is it the notion of bravery that the firefighter faces when entering a burning building that everyone else is running out of? Maybe it's a family thing, born into it; it's what some were meant to do. These reasons may be true for some, but not for others. A number of individuals enter the career just on a whim. Stories have been told by some that they just saw the advertisement in the paper and thought why not? Others were drawn to it because of job security, great benefits, time off and salary. Which ever the reason may be, either childhood dream or job security, everyone who has the privilege of entering such a noble profession doesn't just become a firefighter, they become part of a rich tradition that dates back to the time of the Romans.

Being a firefighter isn't for everyone. It definitely takes a certain type of person, though precisely what that kind is cannot be put into words. Firefighting attracts some of the most unlikely of personalities, bringing individuals together who otherwise would never see the likes of each other in everyday life. Firehouses all across the nation are melting pots of men and women with different religious beliefs, ethnic backgrounds, lifestyles, and political views. Regardless of birthplace or beliefs, those who choose to become a firefighter all become part of what is known as the brotherhood upon entering the fire service.

Aside from reasons of what attracts one to what is believed to be a noble calling, there are two things that will never change. They are history and tradition, ideals that will forever cast a bond among all firefighters regardless of their unique individual characteristics. Whether you have a firefighter that just does the job for the paycheck or the firefighter that does it for the love of it, history and tradition will forever be a constant.

But where did it all begin? What makes the fire service full of history and tradition more than that of any other profession? Everything comes from something, but where? Why are fire engines red and why is the Dalmatian the breed of dog most associated with the fire service? What is the true story of the Maltese cross and where did the paramedic program start and how did the star of life become part of the fire service? The answers to all of these questions and more need to be addressed and separation between fact and fiction need to be told. This is your history. This is your tradition. If only for a brief second that you become aware of the roots that support this great profession will you undeniably embrace the career that is so rich and rewarding.

Where It All Started

Nearly 2 million years ago, Homoerectis inhabited the continent that is now known as South Africa. He possessed an essential human trait, more brainpower than that of his ancestors. He faced many challenges, one, that instead of being atop of the food chain; he found himself at times at the bottom looking into the powerful jaws of his arch nemesis, the leopard. Unless he found a way of protection, he would be the perfect prey, in fact he was. Homoerectis lived in constant fear of attacks. He could not outrun the animal or even withstand a chance in a fight. The one thing that he had that would help him outwit his opponent is that of having a bigger brain. If he could use his brain to control fire, he would have a method to fight back against the fangs. But could he do it?

Burned animal bones have been found from 1 million years ago during the time when Homoerectis walked the earth. These bones were speckled with white marks showing evidence that they were subjected to high, intense heat for over a long period of time. These bones were found in caves where he was known to live, signifying that he had found the ability to control fire. At first, these bones that were found were thought to have been from an animal that was trapped in a natural wild land fire that was probably caused by a lighting strike. Due to the white marks which represented that the bones had been subjected to high heat, this could only mean that it had been caused by a stationary fire. Tests were done to prove that the bones were not charred white by the wild land fires. Conclusion was met when an antelope carcass was placed in the path of a fast moving grass fire. The bones were charred but not to the degree of the bones that were found in the caves were Homoerectis lived. He indeed found a way to control fire, but how?

It is believed that Homoerectis did not have the cognitive ability to start a fire, but would have been capable of taking a twig from a brush fire and that alone would transform his world. He would now have the capability to cook food, heat cold winter dwellings, and most importantly, scare away his enemies. It's probable that no creature has ever done this before. This may seem like an insignificant event compared to what we now know, but in reality, it would prove to be a remarkable technological development. Homoerectis became extinct around 50 thousand years ago. Though he cannot take credit for what we have evolved into today, he is however accredited with the ability to control fire, which has stuck with us in our time.

Now with man being conscious of the power of fire, it has either become friend or foe. It is logical to say that lightning first created fire, but there are many myths relating to the origin of fire. One myth states Prometheus a Greek God, kindled the first fire upon earth, and another which conversely declares that man stole it from the gods, but its true source will never exactly be known.

With the acquisition of their new- found means, came the problem of maintaining it. Interesting examples of creativity were soon presented with possibly first, fire being preserved in the ashes of itself. A few other examples talk about a type of slow-match or fire-stick that was developed, and later, when man worked with metals, the curfew, or "fire cover" was invented.

Now with the ability to preserve fire comes the dilemma of keeping it in its non-destructive form. However this would not be possible. With evolution comes greatness. With greatness a new evil awakens. Innovations with building materials, construction methods, flammable liquids, nature's wrath, and human error would all contribute to fire becoming a terrible beauty that would destroy lives and property from here on out. As a result of the rise of a new enemy comes the creation of the firefighter.

Quote

"If Prometheus was worthy of the wrath of heaven for kindling the first fire upon earth, how ought all the Gods to honor the men who make it their professional business to put it out?"

John Godfrey Saxe, circa 1850

The Romans Contribution to Firefighting

The history of Rome is plagued with records of great fires that nearly destroyed the city from time to time. To combat these fires, Rome organized some of the first fire companies in history. The history of the fighting of fires dates back at least to ancient Egypt where hand-operated pumps may have been employed to extinguish fires. However, such attempts could be of limited value given the possibility of large fires that could sweep through Rome and other cities.

During the early years of the Roman Republic, bands of slaves were employed as firefighters and they would carry around these Egyptian-made advanced pumps. They (the slaves) were called the 'familia publica'. Regrettably, they were inadequately trained, scantily equipped and not very motivated. As a result, there was a lack of firefighting going on so some clever and cruel businessmen, such as Crassus, made fortunes in the "business" of firefighting.

Crassus had his own team of slaves that worked as firefighters who would go to the burning house, and as it burned, Crassus would attempt to buy it. As the building burned more and more, Crassus would offer less and less money. As soon as the owner conceded to sell to Crassus, the slaves would put out the fire.

In the year 6 A.D., almost 30 percent of Rome burned in a very large series of fires. After this devastating loss, the Emperor Augustus passed duty for firefighting to the Roman Army and instituted the Corps of Vigiles who were to protect the city of Rome and it's Empire for the next four hundred years. Augustus's Corps of Vigiles fought fires using bucket brigades and pumps, as well as poles, hooks and even ballistae to tear down buildings in advance of the flames. It is generally thought that this is where the "hook" in "hook and ladder company" comes from.

During this time, only citizens of Rome, those who were considered free at birth could join the Roman Army. However, as the first Vigiles were enlisted from slaves they were never considered true Roman soldiers, and because of this, they carried no banners. Nevertheless, because recruitment was slow, the incentive of full citizenship was offered to free the slaves after six years of service.

After a century, freed men would enter the Corps of Vigiles merely for the prestige that came with it. The Corps of Vigiles was seen as an honorable means of obtaining full Roman citizenship. At one time it was thought that the Vigiles consisted of 7000 paid firefighters. A normal period of service with the Corps lasted 26 years.

The Corps was ranked much like an elite military service. The Vigiles had become so powerful that they could break into a house if they had suspicion there was a fire inside. If a householder was found to be negligent by allowing a fire to start, the commander could have him flogged.

Somewhat fascinatingly, firefighting was not the only function of the Vigiles. At night they would 'police' the cities and towns and quite often be required to track down and recapture runaway slaves. During the day they would be stationed at public baths to watch over the clothes of the bathers to prevent pilfering by the public and bath attendants.

Three ranks have been identified within the Corps of Vigiles.

The Aquarius - This was a firefighter whose main duties were to supply water to the siphons or pumps and the organization of 'bucket chains'.

The Siphonarius - This firefighter was responsible for the supervision and operation of the water pumps.

The Uncinarius - This firefighter was a 'hook' man. He carried a large fire hook for pulling off burning roofs.

Each and every firefighter carried knives, axes and spades to help them in their duties. All this equipment and the training the Romans received was as diverse and extensive as any you may have found in any early 19th century Fire Brigade in Britain. The Romans were remarkably forward in thinking with regard to their fire prevention and firefighting methods.

The Vigiles were successful in fighting fires for over four hundred years. The only major fire in Rome during this time was in 19 July 64 A.D. that started near the Circus Maximus which eventually destroyed two thirds of Rome. Theorists say the Emperor Nero may have forced the Vigiles not to act, so he could clear room for his new palace.

Discrepancies in whether or not Roman Fire Brigades remained solely on Roman soil lie in the following. For historical purposes and with either ending unclear, both conclusions have been incorporated.

- Despite their success in Rome, the *Vigiles* concept did not spread to other cities in the Empire. This was probably due to political factors at the time. Letters exchanged between Pliny the Younger, governor of the province of Bithynia in Turkey, and the Emperor Trajan show that despite local willingness to set up a fire brigade, Rome did not want independent fire brigades in the outer territories.

- The Corps of Vigiles was not confined to Rome. The army list drawn up in AD 410 gives a variety of locations of Vigiles stationed all through the Roman Empire including Britain. However, with Rome in jeopardy, the Legions started retreating to Italy and the Corps of Vigiles, known as a highly regimented force, was armed and used to protect the outposts of the Empire. Vigiles finally disappeared with the collapse of Rome.

Despite either conclusion, it was said that, "Europe was not to know another well-trained and well-equipped Fire Brigade for a thousand years."

Saint Florian

Saint Florian is known to be the protector of firefighters and continues to watch down upon all who take on this noble calling. A third century officer with the Roman army stationed in modern Austria, military administrator of the town of Noricum, and a closet Christian, Florian is said to have stopped a town from burning by throwing a single bucket of water on the blaze. How did such a man become immortilized in fire service history?

Florian was born in Austria in the 4th century about 250 A.D. in Cetium (Austria). The St. Florian commemorated in the Roman Martyrology on May 4th, was an officer of the Roman army. He joined the Roman Army as a youth, and through hard work and determination he advanced through the ranks. He soon occupied a high administrative post in Noricum, now part of Austria. The Emperor Diocletian and his assistant Maximian liked Florian because of his abilities to solve problems and work with people. They saw Florian as a man capable of handling heavy responsibilities. The saint suffered "death for the Faith" in the days of Diocletian. Although he became an officer of the Roman army, he didn't observe a "don't ask, don't tell" policy and confessed to his Christianity in anti-Christian times.

The Emperor sent an assistant named Aquilius to the region controlled by Florian to kill all the Christians in the area, and to find out why Florian was not following orders. When he finally met up with Florian, he asked him why he refused to persecute the Christians to whom Florian replied, "Tell the Emperor that I am a Christian and will suffer the same fate as the Christians." Aquilius then offered him a raise and promotion if he would change his mind. Florian refused. Aquilius was outraged and had his soldiers beat him with whips, yet Florian stood steadfast.

He told Aquilius that he had suffered many wounds for the Emperor, why not a few scratches for his own beliefs? His courage scared Aquilius who feared that Florian would lead others to rebel.

It was for these reasons that the Emperor was shocked to learn that Florian was not following his order to punish all Christians in his area. This punishment included burning Christian churches and books, firing all Christians from their jobs, putting them in prison without good reason, and finally forcing them to sacrifice to the Roman gods. Florian did not carry out his orders to persecute all Christians in the area, and, thus, was sentenced to death by fire.

His legendary "Acts" state that he gave himself up at Lorch to the soldiers of Aquilinus, and the governor of Lorch, when they were rounding up the Christians. Standing on the funeral pyre, Florian is reputed to have challenged the Roman soldiers to light the fire, saying "If you do, I will climb to heaven on the flames." Apprehensive of his words, instead of burning Florian, he was flogged and flayed after making the bold confession. Not willing to take a chance, Aquilius decided to drown Florian by tying a giant stone around his neck and throwing him into the river Enns. A pious woman later recovered his body, and he was buried honorably.

Because of his stance and resistance to death by fire, St. Florian became the protector of firefighters all over the world.

Firefighting in Colonial America

The history of firefighting in America goes all the way back to Jamestown, Virginia, the first permanent English settlement in the new world. Founded in 1607 by colonists from the London Company, Jamestown was under the command of Captain James Smith. It did not take long for fire to unleash its unforgiving fury upon the settlers.

January, 1608 brought what was to be the beginning of change to the new settlement. A destructive fire ruined most of the colonists' supplies and houses bringing to light a terrible realization. Captain Smith made a brief evaluation of the situation stating, "I begin to think that it is safer for me to dwell in the wild Indian country than in this stockade, where fools accidentally discharge their muskets and others burn down their homes at night."

Bright eyed and eager, the population of the new world would continue to rapidly rise as shiploads of immigrants stepped ashore searching for a fresh start in a new land. Cities began to take shape, and the problems Smith found in the small stockade multiplied as more and more structures were built. The fire load in these cities increased as forests were cleared and wooden homes and buildings were erected. Communities that sprang up around three of the best harbors, Boston, New York, and Philadelphia soon faced a number of social problems concerning housing, sanitation, water supply and the danger of fire. These three cities, and the firefighters who would step forward to protect them, set the groundwork that would forever shape the American fire service.

In 1648, New Amsterdam (later to be named New York) Governor Peter Stuyvesant stood confidently on his peg leg and selected four men to act as fire wardens. These men were authorized to inspect all chimneys and to fine any violators of the

laws. The city councilmen soon appointed eight prominent citizens to the "Rattle Watch" these men would patrol the streets at night carrying large wooden rattles. If a fire was seen, the men spun the rattles and then would direct the responding citizens to form bucket brigades. This is generally recognized as the first step in organized firefighting in America.

Even earlier, Boston's city fathers took the first strides in fire prevention when Governor John Winthrop outlawed wooden chimneys and thatched roofs in 1631. Forty years later, Boston experienced a series of arson fires and finally an inferno in 1676. The small "ingine" built by local iron maker Joseph Jynks, had little effect on the growing wall of flames. Shortly after the conflagration, Bostonians sent for the "state of the art fire engine" then being made in England. The three foot long, 18-inch-wide wooden box arrived with carrying handles and a direct force pump that fed a small hose. The tub section of the engine was kept filled with water by a bucket brigade.

The need to coordinate firefighting efforts brought about the organization of the first engine company to be known in colonial America. Twelve men and a captain were hired by the General Court to care for and manage the engine and to be paid for their efforts. On January 27, 1678, the company was put in service. Its captain, Thomas Atkins, was actually the first firefighting officer in the country.

Two Newsham engines arrived in New York in December 1732. Jacob Turck was selected to be in charge of the engines and to keep them in repair at his own cost after a 10-pound salary was advanced him. Turck also worked other ideas. One, a pump of his own design, which was perhaps the first mechanical fire engine built in America and the other was the invention of the first leather fire cap.

Most notable among the famous Americans who helped shape the country and the fire service was Benjamin Franklin, a writer, printer, philosopher, scientist, statesman of the American Revolution and a fireman. Franklin helped draft the Declaration of Independence, served as a diplomat, and invented items that ranged from lightning rods to bifocal eyeglasses.

Franklin often wrote about the dangers of fire and the need for organized fire protection. He was dissatisfied with Boston's Mutual Fire Societies (also known as "Fire Clubs") because the "Fire Clubs" existed solely for the protection of its members, and not the community as a whole. Franklin wanted organizations that would battle all fires, regardless of whose property was burning.

After an extensive fire in Philadelphia in 1736, Franklin established the first *volunteer* fire department, which consisted of 30 courageous citizens who were named the Union Fire Company. They would soon become the standard for volunteer fire company organization. As the idea of volunteer fire brigades gained popularity, additional companies were formed in Philadelphia. Each of the companies paid for their own equipment and located it throughout town at strategic places.

Other famous Americans who served as volunteer firefighters include: George Washington, Thomas Jefferson, Samuel Adams, John Hancock, Paul Revere, Alexander Hamilton, John Jay, John Barry, Aaron Burr, Benedict Arnold, James Buchanan and Millard Fillmore. Volunteer firefighters played a priceless role in protecting lives and property back then and continue today in many communities across the nation.

Two vital tools utilized by early American firemen were the bed key and salvage bags. With firefighting apparatus able to supply only a small stream of water, a fire that began to gain any headway would soon be out of control. Arriving firemen quite often opted for immediate salvage efforts in the fire building and surrounding exposures. The bed key was a small metal tool that would enable the men to rapidly disassemble the wooden frame of a bed, more than likely the most expensive item owned by a family and remove it to safety. Other valuables were also snatched up, placed in salvage bags and carried to safety. This method of firefighting is now compared to salvage performed in the fire service today.

The first attempt at fire insurance went bust after a devastating fire in Charlestown, Massachusetts in 1736. Ben Franklin then organized the "Philadelphia Contributorship" in 1740 to insure houses from loss by fire, a business enterprise that was a success. The company adopted "fire marks" to be fastened to the front of the insured's property for easy identification. These fire marks were to distinguish between insurance companies and the properties they insured.

Now with rules to govern firefighting operations and the equipment such as buckets, hooks, ladders and the formation of volunteer companies, firefighting started to become formalized. The chain of command would soon fall into place as officers of various ranks were established. Firemen devised new and better ways to accomplish their mission with everything from helmets to hoses being invented or improved. Firemen in Philadelphia, New York, Boston and other cities made major advances in the technology and theory of firefighting.

The legacy of colonial firefighters can still be seen in fire department operations and organization across the country to this day. The wooden hydrants, horse drawn steam engines and a mostly defensive fire attack as opposed to a now more offensive attack, are gone, but the iron willed determination of American firefighters is as strong as ever.

THE ENEMY

I am more powerful than the combined armies of the world.

I have destroyed more men, women, and children than all the wars of all the nations. I massacre thousands of people every year.

I am more deadly than bullets and I have wrecked more homes than the mightiest guns.

In the United States alone, I steal over one billion dollars each year.

I spare no one and find my victims among the rich and poor, the young and old and the strong and the weak.

Widows know me to their everlasting sorrow. I loom up in such proportions that I cast my shadow over every field of labor.

I lurk in unseen places and do most of my work silently.

You are warned against me, yet you heed me not.

I am relentless, merciless and cruel.

I am everywhere, in the homes, in the schools, in the factories, on land, on sea, and in the air.

I bring sickness, degradation and death. Yet few seek me out to destroy me.

I crush, I maim, I devastate. I will give you nothing and rob you of all you have.

I AM YOUR WORST ENEMY

I AM RAMPANT FIRE

THE HISTORY

First Fire Engine

The term "fire engine" was possibly first used in the 17th century, in exactly the same sense it is used now, a machine for throwing water to extinguish fires. The evolution of the fire engine spans over a time frame prior to that of the Roman era. There are many accounts of "firsts" when it comes to taking credit for the invention of the fire engine.

Though the Romans have been credited with many technological firsts such as the invention of the roadway systems, establishment of running water in a city through aqueducts, and the organization of some of the first fire companies in history, they cannot take credit for the creation of the fire engine. The first fire engine, as we know it, was not a Roman invention, but an Egyptian one. Technically it would not be an engine but rather the first fire pump. The city of Alexandria, between second and third century B.C., saw the first primitive device that was used to spray water. Ancient inventor Ctesibius made a pump which was later improved by Heron in the first century B.C. Much like a primitive engine, Heron's pump had pistons that were moved by a rocker arm that pivoted from a center post. The pipe that led out of the pump could be moved up and down, or left to right. The remains of such a pump have been found both in Italy and in England.

The fire pump was then reinvented in Europe during the 1500's, reportedly used in Augsburg in 1518 and Nuremberg in 1657. A book of inventions mentions in the year 1655 a steam engine pump called the "fire engine" used to raise a column of water 40 feet, but there was no reference to whether or not it was portable.

In 1654, a local iron maker, Joseph Jynks of Saugus, MA, built a small "ingine" which was probably a syringe-type pump. Boston experienced a series of arson fires and finally an inferno in 1676. The iron maker's pump was ineffective against the advancing flames. In 1679, Boston then sent for what would be considered a "state of the art fire engine", then being made in England. Philadelphia obtained a hand-pumped fire engine in 1719, years after Boston's 1654 model appeared, but before New York's two engines arrived from London.

There are also accounts that the fire engine was perfected by Richard Newsham of London in between 1725 and 1730. The engine would be pulled as a cart to the fire. These manual pumps were manned by teams of men and could deliver up to 160 gallons per minute at up to 120 feet. In 1731 Newsham's models were sent to New York six years before the formation of the NYC volunteer fire department. The first pumping engine built in America was attributed to Thomas Lote. In 1743 the engine was created and named "Old Brass Backs" because of his excessive use of brass trim.

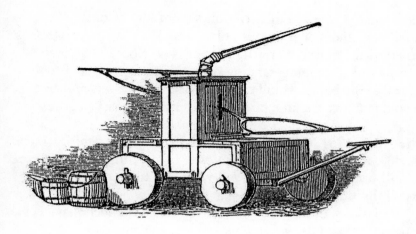

The first true steam engine that was horse-drawn, was developed in London in 1829 by a gentleman named John Braithwaite and was assisted by John Ericsson.

Paul Hodge, who came to America from England sometime before 1837, built the first steam fire engine in United States. On March 27th, 1841 he demonstrated the apparatus named the

"Exterminator" in front of City Hall to firefighters, city councilmen, and many curious onlookers. His new invention was scorned by the volunteer firefighters of New York due to many concerns and complaints ranging from taking too long to fire up to fear that the boiler would explode. The engine was returned and sold to a box manufacturer for use as a stationary engine.

In the fall of 1851, in Cincinnati, a fire broke out in a mill, which brought outside volunteer companies to help extinguish the blaze. The mill was ultimately destroyed when the 13 companies stopped fighting the fire and began to fight amongst themselves. The city fathers, outraged by the repetitiveness of such instances, wanted to disband the volunteers and create a paid fire department. The key to such a possibility would be to find a way to fight fires effectively with fewer men. Because the city was centered for steamboats and locomotive construction, the city fathers, already familiar with steam power, decided to seriously explore the possibilities of steam fire engines. Two entrepreneurs in Cincinnati, Alexander Latta and Abel Shawk had come up with an idea for a fire engine with a steam-powered pump. The new machine nicknamed "Uncle Joe Ross" after its sponsor was demonstrated on January 1st, 1853. It would revolutionize firefighting in more ways than one. On March 10th, 1853, the Cincinnati city council voted to create what would be the first paid fire department in America starting on April 1st of that year.

In the mid to late 1800's horses would be phased in as the primary means of pulling apparatus.

In 1824 the basic principles of the internal combustion engine were developed and soon the fire service would take a technological leap that would be utilized all the way through modern day times.

Other "firsts" worth noting are the first successful aerial ladder and water tower. The aerial ladder was patented in 1868 by Daniel Hayes. It was constructed of wood and required several firefighters to raise the ladder by hand through a series of gears and pulleys. The rights to this patent were eventually sold to LaFrance. The Fire Extinguisher Manufacturing Company, Babcock and Dedcrick, also developed their own aerial ladders.

In 1879, John Hogan and Abner Greenleaf developed the first water tower. This tower consisted of a 50-foot mast of pipe sections that had to be assembled to attain the desired height and was then raised manually by cranks and gears. Water towers were capable of delivering a large-capacity elevated stream into the upper floors of burning structures.

As with steam engines, the first gasoline engines were gradually improved over time. The first gasoline automobiles were seen on the roads in the 1890's, but it wasn't until the year 1904 that fire service saw its first gasoline-powered American-Mercedes touring car that was presented to Chief Edward F. Croker of the New York City Fire Department. Fire engines would now take a step that would, over time, phase out the use of horses to pull the machines. Gasoline-powered motors and pumps began to appear in the fire service and the invention of the first pumper with a single engine to do both driving and pumping occurred. By the 1920's, the disappearance of the steam engines would mark a page in the history books.

Are the Egyptians able to hold rights to the creation of the first fire engine or was Newsham's invention the first or just a modification of earlier versions? No matter where the actual "first" fire engine came from, it did however lay a solid foundation for what we see in today's modern fire engines. The engines of today have great pumping capabilities, upwards of 1500 gallons per minute and have tanks that are able to hold 750 gallons of water or more, a far cry from the small hand tubs that needed to be fed by bucket brigades and literally hand pumped.

So what does the future of fire apparatus design hold? The American fire apparatus industry is in the midst of a restructuring. In recent years, many of the industry's manufacturers have been sold, merged, or gone out of business. This restructuring continues with acquisitions and mergers. The end result is hard to predict, but it's possible that only a handful of major manufacturers or aligned companies will remain.

The trend toward commercial chassis will continue, as will the greater use of foreign chassis on which to mount fire apparatus body work. In all probability, American apparatus will soon mirror European designs to a greater degree. Much of the industry restructuring that is going on involves foreign parent companies that will ultimately look to market their products worldwide. Another trend that is taking hold is the multi use vehicle known as a "Quint." These vehicles are able to perform the functions of an aerial ladder combined with an engine. Alternate-fuel vehicles are sure to come, but like the changeover to horses, gasoline and diesel, will be slow to take hold. Most important, fire apparatus will continue to be interesting, eye-catching vehicles.

Why Fire Engines are Red

The most widely accepted explanation that fire engines are painted red dates back to the 1800s, an era when there was fierce competition between the fire brigades of neighboring cities and towns. The firefighters of each brigade took great pride in their fire apparatus. Each brigade wanted their rig to stand out by being the cleanest, having the most brass, or being a noble color. Because red was the most expensive color, that's what color most crews chose to paint the apparatus. But the question is who pioneered the color and why was it the most expensive?

These early firemen were held in very high regard by their communities and they, in turn, took great pride and group distinctiveness in their engine and its overall appearance. Brass and copper straps and other fixtures originally added for strength became more numerous and decorative. Scrollwork, murals, lanterns and fancy wrought ironwork all became fairly common as the companies competed with one another to make their "masheen" a standout.

Prior to the 1850's, most hand fire engines were painted in greens, blues, or black with white or yellow accents. When red hues were used, they were a rather muddy oxblood or cordovan color, comparable to what you may have seen on the floors of an 18th century home. Since many of the volunteer fire companies desired to decorate their machines to reflect their individual tastes and traditions, the engines often came from the manufacturer unfinished or in a generic gray primer coat.

In 1847, an Englishman named Murdock developed a method to distill sulfide of antimony into a bright scarlet red pigment. The procedure was very monotonous and expensive which, in turn, made the exciting new color very costly and rather rare. This combination proved alluring to firemen. No doubt the first engines done up in bright red with polished brass and gold leaf trim were stunning status symbols. Not to be outdone, companies everywhere followed the trend and over the next few decades red became virtually the standard color of fire apparatus.

Today, just as you have many more choices of colors available to you for your vehicle, so do the fire engine manufacturers, and it is not unusual to see white, yellow, blue, orange, green, or even black fire engines, in addition to red. Some studies hint that colors such as lime-green may be more visible to the public than traditional red, but still the vast majority of fire departments continue to use red fire engines, a color instantly recognized by everyone as that of a fire engine.

History of the Hydrant

Firefighting was around before the first fire hydrant and the idea that would coin the phrase, "getting the wet stuff onto the red stuff". The inventor of the earliest device that we would recognize today as a fire hydrant cannot be told because the evolution of the fire hydrant spans over a period of many years and has been influenced by many people. The first hydrants were used for public water supply from the earliest municipal water systems. They resembled faucets and were at best, suited for the bucket brigade method of firefighting. Prior to municipal water systems, there were other means to provide water in the event of a fire.

In ancient China, firefighting cauldrons were placed in strategic locations around a village or city and kept filled with water as to be ready in the event of a fire. In colonial America, cisterns were used to store water for early fire fighting purposes and these continued to be used even after the introduction of the hydrant in many cities. Moreover, as late as 1861, Louisville, Kentucky employed 124 cisterns but no fire hydrants. These cisterns are still used today for firefighting.

The Advent of Pressurized Municipal Water Supply

The term "fire plug" dates from the time when water mains were made from hollowed out logs. The fire company would respond to a fire, dig up the cobbles down to the water main, then bore a hole into the pipe so that the excavation would fill with water which they could draft using their engine. After the fire was out, they would seal the hole with a "fire plug". The next time there was a fire in the neighborhood, if possible, the same hole would be dug up and the same plug would be used so they would not have to cut into the main again.

This practice was in common use until the year 1666 after a conflagration destroyed three quarters of London, the city installed new mains with predrilled holes and plugs that rose above ground level. In the 1700s, valves began to replace the simple wood stoppers, and firefighters began carrying portable standpipes or vertical outlets, which were inserted into the plugs. This basic configuration is still in wide use in Britain and other European nations today.

Cast iron would eventually replace wooden water mains, and when this type started becoming popular, branched fittings were placed on the mains at intervals, much like today's fire hydrants. These were similar to underground hydrants which could draw water from the water mains in a crude fashion.

The first American style post or pillar type hydrant is attributed to Mr. Frederick Graff Sr., Chief Engineer of the Philadelphia Water Works, around the year 1801. The hydrant had a combination hose/faucet outlet and was of "wet barrel" design with the valve in the top. It is said that Mr. Graff held the first U.S. patent for a fire hydrant, but this cannot be verified since the patent office ironically burned to the ground in 1836, destroying all the U.S. patent records.

Cast Iron Hydrants

In 1802, the first order for cast iron hydrants was placed with cannon maker Foxall & Richards. In 1803, Frederick Graff Sr. introduced an improved version of the fire hydrant with the valve in the lower portion. The valves were inserted into wooden mains with a tapering joint. In 1811, Philadelphia claimed to have 230 wooden hydrant pumps and 185 cast iron fire hydrants.

On the subject of the further improvement of the cast iron hydrant, the R.D. Wood Company's catalog of 1877, gave this statement:

"In 1803 Frederick Graff, Sr., designed for the then recently constructed Philadelphia water-works, a stand-pipe intended to remain permanently in position and to be constantly charged with water ".

"This was a most important advance in the design of fire-plugs, since it gave us a hydrant that in mild climates might remain ready for instant use. Its valve was placed at the bottom of the stand-pipe near the level of the top of the main pipe, and it introduced a drip, or waste, that opened by action of a spring as the main valve closed, so that all water remaining above the main valve in the stand-pipe at once drained off, provided the spring was still in order. This model of hydrant, which was admirable in many respects for use in southern cities, was for many years followed generally in the construction of similar apparatus in other of the larger American cities. The nozzles of these hydrants were generally placed about two feet above the ground surface, so that they might be above obstructions of mud, snow and ice, and they were generally housed by a covering box of iron or wood, that was removable to afford access to the valve-rod key ".

"...The necessity, on account of above faults in principles of construction, of packing and covering hydrants in winter with manure, tanbark, straw, &c., as practiced in many cities, to lessen the liability of freezing". Such was the hydrant, varying but slightly as made by different manufacturers, in general use throughout the United States, when the "Mathew's Hydrant" was first introduced."

In 1865, Philadelphia had installed cast iron hydrants that were very comparable to today's models. According to the first annual report of their public waterworks, the City of Buffalo, NY was still installing wooden case hydrants in 1869, a time that the days of the wooden case hydrant were over.

Many companies were now making cast iron fire hydrants with hydrants being deployed to both major metropolitan cities and smaller less densely populated ones. Europe, too, was installing such hydrants with Zurich, Switzerland having their first hydrant system in place by 1870 and in Asia, the City of Yokohama, Japan, had installed 131 hydrants in 1887.

There are three basic types of hydrants that were established for connection to the pressurized municipal water supply: the *dry barrel*, the *wet barrel* and the *below ground* or *flush type*. The main challenges faced by designers of these hydrants was the potential for freezing, hydraulic efficiency, and ease of repair. These problems were all known and dealt with, to varying degrees of success, early on, before 1900.

Although materials have improved and some of the elements of hydrant design have been refined, the basic form of both the dry barrel and wet barrel hydrants have endured relatively unchanged since the mid 1800s. However there are still many areas today that do not have hydrants. In many rural settings water shuttle operations are the only means of getting water to a burning structure. Fire hydrants are mostly seen in urbanized areas and are a vital part of firefighting operations.

The Advent of the Fire Extinguisher

In about 1660, Ambrose Godfrey Hanckwitz, an apothecary from Hamburg in Germany, was invited to England by Sir Robert Boyle to assist in his laboratories and in the manufacture of chemicals, including the first phosphorus made in England. He dropped the name Hanckwitz when he settled in London, and had his own laboratory built at the corner of Maiden Lane, Covent Garden, where he continued to make significant developments in chemistry.

Godfrey also invented a fire extinguisher. It worked by suffocating the fire and therefore putting it out, an idea credited to a Zachary Greyl. A miniature wooden barrel was filled with fire-extinguishing material. Then gunpowder was inserted in a pewter sphere at the center of the barrel, and fitted with a fuse. When the fuse was fired, the explosion forced the fire upwards. It was first tested on Hampstead Heath in 1723. A report about a spate of fires in London in 1727 said:

"I hear that the famous machines or Fire Watches, invented by Mr. Godfrey the great Chemist...displayed their wonderful effects, and prevented the progress of that furious element [the fire]."

However, Godfrey's extinguisher appears only to have been used for a few years.

It is thought that George William Manby (1765-1854) invented the first fire extinguisher between the years 1813 and 1816. Manby, who had been a member of the British militia, had observed the inability of firemen in Edinburgh to reach the upper floors of burning buildings and was inspired to create a means of remedying that difficulty.

Manby's extinguisher consisted of a four-gallon copper cylinder which held three gallons of water; the remainder contained compressed air. When the stopcock at the top of the cylinder was engaged, the compressed water would be forced out through a tube running from the valve to the inside base of the cylinder and directed toward the fire.

In 1866, Frenchman Francois Carlier invented an extinguisher which had a cylinder containing a mixture of water and bicarbonate of soda and a separate bottle filled with sulfuric acid. When the bottle was punctured, the acid mixed with the bicarbonate, producing carbonic acid which, when it bubbled, forced the water out.

Another type of extinguisher invented in the late 1800's was a glass grenade-type.

These grenades were made of clear or colored glass usually having a fancy cut-glass, round ball-shaped body, with a tall neck and contained a harmless salt water and brine solution. They were thrown at the fire extinguishing the flames.

A Russian, Alexander Laurent, developed a solution of aluminum sulfate and bicarbonate of soda in 1905. The carbonic acid bubbles, rather than water, were forced out of the cylinder, smothering the fire. Laurent's device was meant to be used on oil-based and electrical fires, against which water is useless.

Around 1912 Pyrene pioneered the carbon tetrachloride or CTC extinguisher. The liquid was expelled from a brass or chrome container by handpump, usually of 1 imperial quart or 1 imperial pint capacity but also made in up to 2 imperial gallon size, onto a fire. The CTC vapoorized and extinguished the flames by chemical reaction. This extinguisher was suitable for liquid and electrical fires and was popular in motor vehicles for the next 60 years. The vapour and combustion by-products were highly toxic and deaths did occur from using these extinguishers in confined spaces.

During World War II, a commercial product called Aero Foam, a derivative of soy protein, was invented by Percy L. Julian and used by the United States military. Modern fire extinguishers have an inner cartridge containing carbon dioxide that acts as a pressurizing agent. When the operating valve is pressed, the gas is released into the main cylinder and forces the extinguishing agent, whether water, foam, or powder, through a nozzle.

Other possible inventors of the fire extinguisher where patents were granted are William Van Deursen in 1870, Finley Latta and John Steiner in 1873. Again with many possibilites and variations of what has come to be known today as the modern fire extinguisher, we may never be sure who extactly was the first, but we can still reap the benefits of a device that saves lives and property on a daily basis.

The Development of Fire Hose

Before the advent of fire hose, the earliest firefighting operations in America saw lines of buckets being passed back and forth between a water source and the structure on fire. This method was the best available at the time, but was both labor intensive and very unsuccessful. Large amounts of water were lost as the heavy buckets passed from hand to hand and the actual delivery of the water to the seat of the fire was poor, unreliable, and dangerous.

With the arrival of hand pumping engines, the ability of water delivery to the fire area was greatly enhanced. Gooseneck delivery nozzles on top of the engines did allow streams to be directed with some precision into the burning building, but the rudimentary machines still had to be filled by bucket brigades.

In Holland, the superintendent of the Fire Brigade, Jan van der Heiden, and his son Nicholaas took firefighting to the next level with the creation of the first fire hose in 1673. These 50-foot lengths of leather were "sewn together like a boot leg." Even with the limitations of pressure, the attachment of the hose to the gooseneck nozzle allowed closer attacks and more accurate water application. Van der Heiden was also credited with an early version of a suction hose using wire to keep it rigid.

Two American firemen made the next major advance in fire hose in 1807 from Philadelphia's Hose Company 1. James Sellars and Abraham Pennock experimented by using metal rivets instead of stitching to bind the seams of leather hose, their efforts proved successful.

Leather hose still had many disadvantages including drying out, cracking and bursting from excessive pressure. The introduction of rivets, however, solved some of these problems allowing higher pressures and greater delivery of water. The improved hose now was 40 to 50 feet in length and weighed more than 85 pounds with the couplings.

This improvement provoked the further development of suction hose to draft larger amounts of water much more quickly than before. The water could be delivered directly to the pump through a hose, thus eliminating the need for buckets. It was said that 100 feet of hose was the equivalent of 60 men with buckets. Hose oilers were developed to keep the leather supple and pliable. Various types of oils and other substances were used to keep the hose in shape. The next upgrade came in 1821, when James Boyd received a patent for rubber-lined, cotton-webbed fire hose. By 1871, the Cincinnati Fire Department was using the B.F. Goodrich Company's new rubber hose reinforced with cotton ply.

With different makers and different sizes of hose finding its way to the fireground, the problem of linking one type of hose to another began to develop. This was officially addressed for the first time in 1873, when the International Association of Fire Engineers held its first convention and adopted a standard coupling size of 722 threads to the inch.

In 1878, the American Fire Hose Manufacturing Company of Chelsea, MA, advertised that it was manufacturing the "first seamless cotton fire hose produced for steam fire engines." Numerous other companies were developing similar hose with pressures tested to 350 psi. Even with the new lighter-weight hose, some fire departments were slow to desert their tried-and-true leather fire hose.

Improvements were made and woven cotton became the standard in the fire hose industry. Better weaves made hose stronger, yet easier to roll and handle. As technology advanced, manufacturers made larger-diameter hoses capable of greater and greater pressures. Small attack lines became stronger, more reliable and easier to handle. Lightweight, durable and flexible fire hose is now common. One can only guess how amazed the early hose pioneers would be with the size, materials used, pressures attained, and the amount of water delivered by today's modern fire hose.

History of the Leather Helmet

Jacob Turck of New York City is credited with inventing the first "fire cap" around 1740. It was round with a high crown and narrow rim and was made of leather. Improvements on his design were made by Mathew DuBois, who sewed iron wire into the edge of the brim to give the helmet shape and strength, and provide resistance to heat, moisture, and warping. The leather helmet as it is known today came from a very modest and non-fire related beginning. Although the year the traditional fire helmet was invented is mired in speculation and debate, it is generally agreed upon as sometime between 1821 and 1836.

The gentleman credited with its founding was named Henry T. Gratacap. Gratacap was a volunteer fireman in New York City, but made his living as a luggage maker. He had made quite a name for himself because of his innovative luggage specifically designed for ocean transit. It was made of leather that was specially treated, which offered unparalleled durability and withstood wetness without rotting. These qualities were very desirable in a fire helmet as well and Gratacap designed the first "eight comb" (a design composed of eight segments) fire helmet. This helmet was originally adopted by the New York City Fire Department (FDNY) in the late 1800s, which

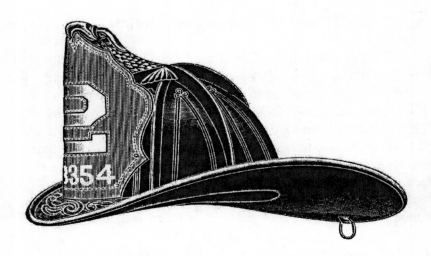

has remained virtually unchanged through approximately 168 years of faithful and steadfast service. This helmet retains the same look and quality that generations after generations of firefighters have relied upon. They are made of stout tanned Western cowhide, a quarter of an inch thick, reinforced with leather strips which rise like Gothic arches inside the crown. The long duckbill, or beavertail, which sticks out at the rear, is to keep water from running down firemen's necks. Originally, these helmets were sometimes worn backwards so the beavertail would protect its wearer from the intense heat of firefighting. Additionally, some tillermen (a name for the driver of the rear section of a tractor drawn aerial truck) would also wear them backwards to protect their faces from rain and snow.

Cairns & Brother's eventually took over after Gratacap's retirement sometime in the 1850s, and the Cairns & Brother legacy was born. They have pioneered firefighter helmet technology ever since. Cairns & Brother's commitment to protecting lives is evident in their "systems," where engineered components synergistically work together for unparalleled protection in harsh environments. As the original OSHA compliant leather helmet, it is individually hand shaped, hand trimmed, and hand stitched to meet the strenuous demands of today's most dangerous profession – firefighting.

Although not a required component of the helmet, some brandish a brass eagle adornment that graces the top of the helmet and secures its front piece. The eagle, it seems, just happened, and has no particular significance at all. Long ago, around 1825, an unknown sculptor did a commemorative figure for the grave of a volunteer fireman. It depicts the hero issuing from the flames, his trumpet in one hand, a sleeping baby in the other, and on his helmet, an eagle. Firefighters were not wearing eagles at the time; it was a flight of pure fancy on the sculptor's part. Soon after when firemen saw it, they thought it was a splendid idea and it was widely adopted.

Today there are many variations of structural firefighting helmets. Different styles, colors, materials and configurations can be seen all over the world and across the U.S., however, the true traditional style which is common today even though it may not be leather, resembles the original "New Yorker".

The Development of Breathing Apparatus

Early firefighters around the world had to face many hardships as compared to what we know as firefighting today. Not only did they have to endure fire and the effects of heat with little or no water supply, but also the debilitating effects of thick choking smoke with nothing at all to protect them. They could not successfully function under the heavy smoke conditions encountered during fires. Something had to change, but how, and who would step up to the plate and come up with an effective solution that would evolve into what we know today as the self contained breathing apparatus?

Fire service legends recount the practice of firemen growing long beards to help them breathe heavy smoke. The theory was a fireman would dip his whiskers in a pail of water, then clinch his wet beard between his teeth and breathe through his mouth, using it as a filter. Pictures of early firemen show many with these long beards.

As a sensible answer to overcome the difficulty of breathing smoke, many seemingly bizarre and odd inventions were attempted. The means of overall success of an invention would come about by determining how the firefighter would breathe through an apparatus. It would either have to be carried with them in a canister, filtered through a mask or be attached to a hose to the air supply outside of the building.

One of the earliest recorded efforts was in France, where the "Apparatus Aldini" was tested in 1825. This was a thick mask of asbestos worn over the head. Another mask made of woven iron wire was placed over the first. The device provided a small margin of heat protection, provided the wearer was able to maintain the air space between the two masks and not allow the iron mask to touch the inner mask. It is thought this mask afforded the wearer only the small amount of trapped air within for breathing. The functionality of the mask left much to be desired, but the scientific testing by Aldini was pioneering. He conducted tests of his apparatus under actual fire conditions. This was the beginning of serious efforts to protect firefighters from smoke as they worked at fires. Many strange and unusual looking devices were invented, manufactured and tried on the fireground with a wide range of results and effectiveness.

Two years before Aldini's apparatus was in use, a patent was issued to Charles Anthony Deane for a smoke and diving device. This system consisted of a closed helmet, flexible air tubes and a pump. The air was directed across the glasses in the front of the helmet to clear breathing condensation. This apparatus was apparently used with some success by firefighters in London and Paris.

In 1824, a miner named John Roberts came up with a "smoke respirator" or hood, which would allow a person "to enter a dense smoke condition without any danger." A variety of types of filter masks were developed and used by firemen in Europe and the United States. In 1861, an inventor named Bradbrooke devised a "smoke and noxious vapor respirator" designed to allow a firefighter to "enter a building however dense the smoke or vapor might be without injury."

The superintendent of the London Fire Brigade, James Braidwood, invented another type of hose mask at about the same time. To supply air and protect the firefighter from smoke, a tube was connected to an air pump attached to the engine outside the fire building. A bulky leather dress and hood were worn to protect the wearer from heat and flames. Thickly glazed eye-holes were provided in the hood. To provide light, a reflecting lantern was worn on the chest. A high-pitched whistle was attached to the hood for emergency communications.

Braidwood tested his invention under harsh conditions during experimental fires in the vaults of the Fire Brigade Headquarters in Wattling Street. The apparatus was used during the rescue of three small children from a burning house on Fetter Lane.

Numerous men and women were also reportedly saved at other fires by men so equipped.

In 1863, a patent was granted to A. Lacour for his invention, the "improved respiring apparatus." This was actually a self-contained breathing apparatus of sorts and comprised of an airtight bag made of two thicknesses of canvas, separated by a lining of India rubber. The device was worn on the fireman's back and held in place by two shoulder straps and a belt around the waist. The bag was filled with pure air inflated with a pair of bellows, and came in different sizes for air durations of 10 to 30 minutes. From the upper part of the bag two India rubber tubes were connected to a mouthpiece that was held in place by biting down with the teeth. Corks were placed in the mouthpiece when the bag was being filled through a faucet at the bottom of the bag. The corks were then removed when the wearer was ready to begin breathing the stored air. It came with a pair of goggles to protect the eyes from smoke, a rubber clamp for the nose and an air whistle that could be pressed by hand to signal for help. Tests made by a variety of fire departments, including New York City, Brooklyn and even the U.S. Navy, proved the device worked to some extent.

The 1870's introduced "Neally's Smoke Excluding Mask", which fire departments were buying and using. This filter-type mask had a small bag of water that was suspended by a neck strap. Connected to the water bag were two sponge filters that were kept wet when the bag was squeezed.

Air was drawn through the filters to the mouthpiece in the facemask. This "most perfect apparatus" was promoted to fire departments for a mere $15.

A portable breathing apparatus designed for work in mines was introduced at a competition being held in the Belgium Academy of Science in 1853. These oxygen rebreathers continued to be improved slowly by a number of people. Bernhard Draeger designed a closed-circuit rebreather in 1903. These units were used for many years in many major fire departments in Europe and America.

The first successful American self-contained breathing apparatus was the Gibbs. Experiments with this unit began in 1915 and by 1918 they were being manufactured by Edison Laboratories in Orange, New Jersey. In 1920, filter masks took a big step forward when Johns Hopkins University and the University of California completed their research on a gas mask designed to be used in a carbon monoxide-filled atmosphere. Their efforts produced a catalyst called Hopcalite that did not absorb or remove the carbon monoxide, but rather oxidized (burned) it and formed the relatively harmless carbon dioxide. This was one of the most important benefits science had given firefighters to that time.

Near the end of World War II, Scott Aviation was manufacturing breathing equipment that permitted air crews to operate at extreme altitudes. One story goes that a number of Scott engineers watched a smoky fire being fought in a nearby building. They were astounded that the firemen had to operate in such a severe smoke condition and they decided to see if they could adapt their equipment to suit firefighting. Working with the Boston and New York City fire departments, Scott introduced the AirPac in late 1945 after a year of field testing.

This basic design was modified and improved as wartime invention gave way to space technology. NASA and its space program became a new testing ground that directly improved work on the fireground. Modern firefighters now have more air, with less weight and a lower profile. Numerous manufacturers currently offer strong, lightweight air cylinders and breathing apparatus with integrated personal alarms and radio systems. Firefighters have come a long way from breathing through their wet whiskers, or the bulky leather dress and hood.

Turnout and Bunker Gear

Turnout or bunker gear worn by firefighters has come a long way since the last century. Starting off as very basic means by only providing warmth and water shielding characteristics with minimal heat protection, the turnout gear worn by today's modern firefighters are leaps and bounds ahead of what was worn in yesteryear.

Initially the practice of the fighting of fires was done from the outside, so exposures to the elements were of primary concerns. As time went on and with the advent of the breathing apparatus, interior fire fighting operations came into play, which would now require safer protective gear.

The term "turnouts" or "turnout coats" was coined as the need to quickly get dressed and to respond to the alarm. Turnout time is the time beginning when the units acknowledged notification of the emergency, to the beginning point of response time. The quicker a firefighter can get dressed means a faster turnout time. Early on when the whole idea of fire fighting was being formed, firefighters wore Civil War-style uniforms that featured heavy wool trousers, a cotton or wool shirt, and a heavy wool tunic. Wool was the obvious choice, because of its ability to guard against heat and cold, and because of its water and flame resistance. Rubber slickers were sometimes worn over the wool uniforms. Firefighters brought their own gloves to the job, usually standard leather workingmen's gloves that offered little or no protection at all. Knee-high leather boots worn in the early years eventually gave way to rubber boots, some of which could be extended to the hips like modern waders (called "three-quarter boots").

An earlier version of what would soon become known as "bunker gear" came from firemen in the mid to late 1830's. In order to ensure speedy turnout times, before they went to sleep they would place their trouser legs over their boot-tops, so that when the alarm was heard, all they had to do was to put on their boots and pull up their pants.

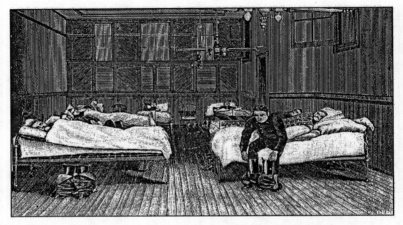

Because they were sleeping in "bunk rooms" at the time, this is where the term bunker gear may have originated.

Turnout gear took a great leap forward after World War II, when various organizations, foremost among them the National Fire Protection Association (NFPA), began issuing standards. The standard on protective clothing for structural fire fighting, for instance, called for an outer layer of flame-resistant fabric that would not be destroyed through charring, separating, or melting when exposed to 500°F for a five-minute period; a second layer to prevent moisture from penetrating through to the wearer; and a third layer to provide thermal insulation from radiant, conducted, and convective heat. Similar standards required gloves that could withstand flame, heat, vapor, liquids, and sharp objects, and footwear resistant to puncture, flame, and heat, abrasion, and electrical current.

In the 1980s, firefighters began wearing turnouts made of three advanced materials: an outer shell material that raised the fire resistance level to about 1,200°F before the material began to break down; a layer that allowed the fire fighter to release moisture from inside the gear; and a fire-resistant synthetic material. Ideally, the latter will last about seven seconds in a flashover situation (when all combustible materials, including walls and floors, suddenly ignite) before catching on fire, which is usually enough time for a firefighter to bail out of room. Further, it is self-extinguishing, meaning once out of contact with a fire, it will not continue to burn. These materials have become the standard for virtually all American fire departments.

Turnout gear soon evolved from a coat and three-quarter boots to a coat and "bunker" pants which is now referred to as bunker gear. The expression "bunker" pants is also said to have been derived from similar pants that were worn by soldiers as they fought in bunkers. These pants would shield the legs of the men protecting them from the blast that came from cannons or guns being fired from the bunker. Bunker pants were slowly introduced into the fire service with much criticism and doubt. As always, there was resistance by firefighters when it came time to switch from the old to the new. Some firefighters would wear the traditional three-quarter boots during the day, and wear the bunker pants at night to speed up once again their turnout times. The new bunker gear is safer and offers the ability for the firefighter to withstand the high effects of heat that they encounter inside while performing search and rescue and extinguishment operations.

Today the lingo is used interchangeably when referring to the protective clothing worn by firefighters. Some call it turnout gear, while others call it bunker gear, while still other refers to the coats as turnout coats and the pants as bunker pants. With many companies testing new materials and technological advancements, who really can imagine what the future may bring for the firefighter of tomorrow.

THE TRADITION

The Maltese Cross

The following story of the origins of the Maltese Cross is a historic depiction of the Knights of Saint John during the crusades while battling the Saracens for control of the Holy Land. Though the romanticism behind the story that surrounds the basis for the modern day Maltese Cross abounds, there have never been any published theories about who first acknowledged the Maltese Cross as a significant symbol for the fire service. It appears that the New York City Fire Department or FDNY was the first department in the United States to adopt the Maltese Cross, but there is no documentation on why it was chosen or who proposed its use. One speculation suggests that the idea of the Maltese Cross for the FDNY was not initially based on any connection to its use by the Knights of Saint John.

The story goes as follows:

Fire Fighters are identified by and are very proud of the badge that identifies them with their department. The majority of fire fighter's badges are in the shape of the Maltese Cross. Why the Maltese Cross? The Maltese Cross is a symbol of protection and a badge of honor with a story that is hundreds of years old.

A courageous band of Crusaders from Malta, known as the knights of St. John, fought the Saracens for possession of the

Holy Land. The need for an identifiable emblem for knights had become crucial. Because of the extensive armor which covered their entire bodies and faces, the knights were unable to distinguish friend from foe in battle. They chose the cross of Calvary as their symbol, since they fought their battles as a holy cause. The cross was later called the "Maltese Cross" and represented the principles of charity, loyalty, chivalry, gallantry, generosity to friend and foe, protection of the weak, and dexterity in service. The Knights encountered a new weapon unknown to European warriors. It was a simple but horrible device of war; it wrought excruciating pain and agonizing death upon the brave fighters for the Cross. The Saracen's weapon was fire.

As the Crusaders advanced over the walls of the city glass bombs containing naphtha struck them. When they became saturated with the highly flammable liquid, the Saracen's hurled a flaming tree into their midst. Hundreds of Knights were burned alive. Others were called upon to perform heroic deeds by rescuing fellow knights and extinguishing fires. In acknowledgement of these feats, the cross worn by these knights was decorated and inscribed. This was considered a most honorable acclaim. Thus these men became our first fire fighters and the first of a long list of courageous fire fighters.

The Maltese Cross is a symbol of protection. It means that the fire fighter who wears this cross is willing to lay down his/her life for you, just as crusaders sacrificed their lives for their fellow man so many years ago. The Maltese Cross is a fire fighter's badge of honor, signifying that he or she works in courage....a ladder-rung away from death.

There are many different versions and conflicting reports of exactly what the Maltese Cross looked like. The "true" Maltese

Cross worn by the knights utilized an eight-pointed cross
resembling four arrowheads converging in the center.

There
were eight points or "tenets" on the cross with each one
representing the knightly merits of:

- Tact
- Gallantry
- Dexterity
- Observation
- Perseverance
- Loyalty
- Explicitness
- Sympathy

In 1259, a declaration by Pope Alexander IV acknowledged the
cross as the Knights of Saint John's symbol and also identified
their battle surcoats of red. However, what we call the Maltese

Cross more directly bears a resemblance to a Cross Pattee. This misuse of language is not limited to the fire service. For instance, the Royal Warrant for the institution of England's Victoria (equivalent to the U.S. Medal of Honor) states that it "shall consist of a Maltese Cross of bronze," but there is without a doubt that it too is a Cross Pattee.

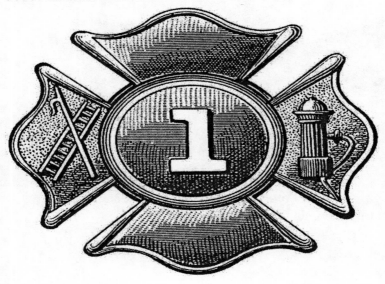

It is fascinating to make note of the significance that the London Fire Brigade (LFB) has sited on its version of the Maltese Cross. What is most unusual is to look at the shape of the LFB badge and compare it to the FDNY badge. To say that both are based on the Maltese Cross raises questions about the soundness of their source. This may suggest that the connection between the fire service cross and the Maltese Cross is based more on the likeness between the ideology of the fire service and of the Knights of Malta than on any similarity in the symbols themselves. Quite a few fire departments in Canada do use a badge that is, beyond a shadow of a doubt, a Maltese Cross. It is modified by a center disk displaying the city seal and department name and sunrays between the arms of the cross or arrowheads.

During the Civil War, two regiments of Union troops consisted exclusively of New York City Firemen. Officially they were the 11th and 73rd New York State Volunteer (Infantry) Regiments, but came to be known as the First and Second Fire Zouaves.

In 1863, General Hooker introduced distinguishing insignias for each Corps of the Union Army. They would be worn on the hats of infantry soldiers and on the left breast by officers.

The symbol of the Fifth and Nineteenth Corps was a Cross Pattee. The Army's adaptation of this cross had flat sides identical to that of the first "Maltese Cross" when it was introduced by the NYFD in 1865 as a hat badge.

The recruiter of New York firemen into its own regiment was Colonel Elmer Ellsworth. Soon after the development of the First New York Fire Zouaves, Ellsworth became the first Union officer to be killed in the war. Soon after, a new regiment,

Ellsworth's Avengers (44th New York Volunteers) was organized whose symbol was also the Cross Pattee, commonly called the Maltese Cross.

So the question still remains, who introduced the Maltese Cross into the FDNY which has now become the universal symbol for firefighters across the U.S.? There are no records to address this, one or more of four individuals may have played a role.

John A. Cregier was an Assistant Chief Engineer in the then NYFD when he enlisted in the 11th New York Volunteers during the Civil War. He served as the third in command under Colonel. Ellsworth, with the rank of Major. Cregier's military service was short when he resigned in 1861 and returned to his position with the NYFD. Cregier served on a committee to make recommendations about the newly formed paid fire department. Is it possible that the proposal for the Maltese Cross badge style, the emblem worn by *Ellsworth's Avengers*, was made by him? It is not known if any other prominent members of the NYFD served in Ellsworth's regiments or had an impact on choosing the Maltese Cross.

Another possibility suggests General Alexander Shaler who was linked with the Army and its use of the Maltese Cross with the Metropolitan Fire Department may have been responsible for establishing its use. He was also recognized with bringing the military format which included the use of military ranks, to the fire department during his term on the Board of Commissioners. General Shaler commanded several different Corps including the Nineteenth whose symbol was the Maltese Cross. There is no apparent reason why he might be partial to this symbol because he was only in command of the Nineteenth for less than six months. Also, Shaler was selected to be on the Board of

Commissioners in 1867, a full two years after the Maltese Cross was implemented by the department. Actually, he didn't leave the military until August 24, 1865, five months after the Act to create the Metropolitan Fire Department was passed and around the same time the actual transition from the volunteer department to paid department began. For that reason it is improbable, though not unlikely, that he played a role in the selection of the Maltese Cross.

Lastly, perchance it was Protection Engine Company 5 that instigated the use of the Maltese Cross. One badge used by the company was a Maltese Cross of an uncertain date. As a volunteer company, it is extremely likely that this badge was in effect earlier to the formation of the paid Department. Nevertheless, how would the badge used by Protection Engine Company 5 be the groundwork for the badge of the paid Department? The answer may perhaps lie with the members of the company. Among them was Foreman Murray Ditchet, a police Captain who first suggested the issuance of badges to firefighters in 1855 and Joshua G. Abbe, one of the original Commissioners of the paid Department, whose own badge was not a Maltese Cross. It may possibly be though that the Maltese Cross badge used by NYFD Engine 5 was the first Maltese Cross used by a fire company.

So, who can take credit for the suggestion of the Maltese Cross's use in the fire service? Unfortunately this question may forever be unanswered. Was its selection based on its use by the U.S. Army or was it the Knights of Saint John? Regardless of its origin it will be an everlasting symbol of the fire service.

Fire Department Rank Insignia

The fire service today is considered a paramilitary organization with much of its structure mirroring the armed forces. For the most part, terminology and the rank structure in the fire service are generally accepted as an idea that came from the Civil War. Prior to the Civil War, U. S. fire departments were often run like a business. Many early fire brigades were private organizations who were paid by insurance companies and therefore had bosses, leaders and foremen and such. During the Civil War entire fire departments joined up as a unit and became infantry units. Many of the elite Zouave brigades were made up of firefighters. General Alexander Shaler, who sat on the Board of Commissioners of the Metropolitan Fire Department (now recognized as FDNY), was with the army during the Civil War. He not only has a possible connection with the use of the Maltese Cross and the Metropolitan Fire Department, but also was recognized with bringing the military format which included the use of military ranks to the fire service.

The leaders during the time of the Civil War were given rank or title, which survived the war. After the fighting most of the men returned to their previous occupation being that of a firefighter. Some men who were captains in the military were known as "Captain" upon returning to the fire department. When they were replaced either through retirement or by death, the department got a new "Captain" for that company.

There were also other influences on the rank structure of fire departments. Possible credit for the first "professional" fire brigade was given to Napoleon Bonaparte. As French emperor, he ordered that a division of the French army known as Sapeurs-Pompiers be used to protect Paris with 30 powerful manual fire pumps by 1800. We can presume that this organization had ranks. In fact many military words, such as those for rank, come directly from old French. (i.e. "Lieutenant" & "Captain") The very first, recorded, full time, fire brigade was in first or second century Rome. This consisted of approximately 7000 "Vigiles". The rank insignia for the fire service is a combination of traditional fire service icons and the common military symbol for such a rank. For many ranks you might see either insignia or both. A captain may have one type on his collar and another on his helmet.

In the early days of North American fire departments, officers, through the use of a large brass device that resembles a megaphone, gave orders to the volunteers. These were very

ornate brass horns.
They were commonly called "bugles" or speaking trumpets. The person with the bugle hanging from his neck was easily identified as the person in charge. Officers became identified with these objects so a small pin in the shape of a bugle became a type of rank insignia for officers. The more "bugles on his collar" the higher the rank, an expression still used today.

Nowadays, fire departments ranks and insignias are still in use. Though, there may be alterations in the ranking system depending on the fire department and their particular organizational structure. Some fire departments or districts may only have one or the other, either a lieutenant or captain. A few departments, depending on their size or politics will only have a chief and no assistant or deputy chiefs. A number of departments carry the rank of engineer and are considered just a step below a lieutenant or captain. No matter the structure, the history of the ranks and insignias are all the same.

Here are a few ways ranks and insignias are used in the fire service today:

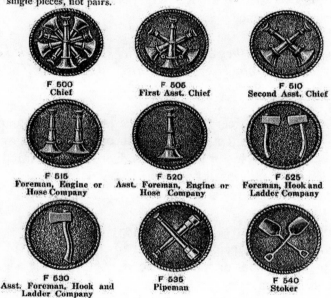

OFFICERS' METAL COLLAR INSIGNIA
These are either gold-plated or nickel, with the insignia in bold relief. Solid metal back with screw or pin fastening. Prices are given for single pieces, not pairs.

F 500
Chief

F 505
First Asst. Chief

F 510
Second Asst. Chief

F 515
Foreman, Engine or Hose Company

F 520
Asst. Foreman, Engine or Hose Company

F 525
Foreman, Hook and Ladder Company

F 530
Asst. Foreman, Hook and Ladder Company

F 535
Pipeman

F 540
Stoker

Firefighter- No particular insignia, but may have a generalized fire department emblem.

Driver - May have no insignia or may have a simple Maltese Cross or something that looks like a steering wheel. They may also have simple "FD" or "FEO" or "Driver" collar brass.

Lieutenant -May have a single bar as in the military or may have a single "bugle" or both. The bar can be gold or silver. Bugle rank insignia can be mounted on a disk or can be just the bugles by themselves.

Captain -May have two bars, just like the US military or may have two vertical bugles. Sometimes, by mistake, you may see a captain wearing two crossed bugles. The bars are often gold but can be silver in appearance.

Note: Depending on whether or not the officers were assigned to an engine or truck, he may have "axes" instead of bugles if he were on a truck company.

Battalion Chief -This officer may be denoted by the traditional symbol of two crossed bugles. Three to four crossed bugles may be used depending upon his or her distance from the chief. If leaves are used, it would possibly be one gold oak leaf (like a major in the military) or one star.

District Chiefs -These ranks usually only exist in the larger departments. Their rank insignia could be any combination of the above stars and bugles or can even be an eagle similar to a colonel's rank insignia.

Note: Some smaller departments may refer to their only on duty chief as District, Deputy, or Assistant Chiefs.

Assistant or Deputy Chiefs-Most often these officers will be wearing bugles or stars. These will likely be wearing three or more of what ever has been deemed the identifier.

It should be noted that in some departments, bugles are what lower chiefs wear and stars are reserved just for the top level assistant chiefs or the other way around.

Chief of the Department -Chiefs either wear FIVE STARS or FIVE BUGLES or both. In order to be able to do this his lower chiefs must have some insignia that suggests it. If none of the ranks under him are wearing more than one star it might not be appropriate for him to have five. Departments try to divide up the stars and bugles so it makes sense. This is why battalion chiefs have such a huge variation in rank insignia. Sometimes a battalion chief is the second highest rank in the department. Sometimes it is the sixth highest rank. The number of bugles or stars often depends upon how far you are from the chief but will never be more than five.

The True Story of Dalmatians

One of the most beloved icons associated with the fire service is the Dalmatian dog. Oddly enough, Dalmatians were the minority when it came to firehouse mascots; the "mutt" was a more popular choice among other pure breads. Still the Dalmatian is and always will be linked with the fire service and its history so it is important to separate fact from fiction regarding this beloved pet.

It is unknown when the breed was introduced to this country, but the first Dalmatian registered by the AKC was in 1888. Some historians believe that the Dalmatian breed originated in Dalmatia, a section of Yugoslavia that was once part of Austria, while others will argue that the dogs derived from India as a lightly spotted hound type dog treasured by gypsies because of their flashy markings and eagerness to have fun. It is supposed that Dalmatians traveled throughout Europe and Asia with the gypsies eventually arriving in Europe around the late 1780's when members of the British upper class brought them home from their frequent trips to continental Europe. However these spotted dogs were also reported in France and Australia and mention has also been discovered in Greek & Roman writings, so it is difficult to say exactly where these gorgeous black and white dogs first originated.

Upon arrival to Britain, the aristocracy hoped to use Dalmatians as hunting dogs but soon discovered that they were better companions to horse drawn carriages because they liked running along side the horses, which also explains why they were called carriage dogs. It is stated that because of its poor hunting abilities, they were relegated to the stable area of fine homes. It was within these stables, that the Dalmatian once again became acquainted with the horses. Though, there are a few references, which have contradicted their hunting abilities, citing that they would hunt along the sides of horses. They were even compared to Labradors.

Before the invention of automobiles, Dalmatians were known as coach dogs. Back in England these dogs would sleep together with horses in the stables, protecting the horses and during the day they would run next to the coach and horses clearing the way. When horse numbers grew here in the new world, so did the number of the dogs and for the same reason they were popular in the old country. Dalmatians were soon adopted by the fire service in the days of the horse-drawn fire wagons because they were agile and not afraid of the horses.

With its superior agility and stamina they could run out in front of the horses and clear the streets for the approaching fire wagon. In addition to guiding the horses, the dogs would keep them company during their long, boring waits between fires. They also served as watchdogs while at the firehouse and while at the scene of a fire. The dogs would guard the equipment against would-be thieves.

When the horses were replaced by gasoline-driven fire engines, many fire departments kept their mascots. When an alarm would sound the dogs would take off and run along side of the pumpers. In some areas you can still see the Dalmatian standing proudly on top of the fire engine as it races to another emergency. The horses are gone from the fire stations today, but the Dalmatians are not. The tradition has been carried on, and it may be as much for the looks and appeal of these beautiful dogs as it is for their nostalgic tie to yesteryear.

Although the above tale may hold some truth, it however isn't exact in its entirety. The fire service has many different breeds of dogs as mascots that served the same purpose, to guide the horses and protect the equipment. The most common dog seen around firehouses were mutts. The Chicago Fire Department has records of some 291 firehouse dogs, mostly being that of a mutt. Nonetheless, the Dalmatian will always be known as the firehouse mascot.

THE TRAGEDY

I WISH YOU COULD...

*I wish you could see the sadness of a business man as his
livelihood goes up in flames or that family returning home,
only to find their house and belongings damaged or destroyed.*

*I wish you could know what it is to search a burning bedroom
for trapped children, flames rolling above your head,
your palms and knees burning as you crawl, the floor
sagging under your weight as the kitchen beneath you burns.*

*I wish you could comprehend a wife's horror at 3 a.m.
as I check her husband of forty years for a pulse and find none.
I start CPR anyway, hoping against hope to bring him back,
knowing intuitively it is too late. But wanting his wife
and family to know everything possible was done.*

*I wish you could know the unique smell of burning insulation,
the taste of soot-filled mucus, the feeling of intense heat
through your turnout gear, the sound of flames crackling,
the eeriness of being able to see absolutely nothing in dense
smoke - sensations that I have become too familiar with.*

*I wish you could understand how it feels to go to school
in the morning after having spent most of the night
hot and soaking wet at a multiple alarm fire.*

*I wish you could read my mind as I respond to a building fire,
"Is this a false alarm or a working breathing fire? How is the
building constructed? What hazards await me? Is anyone
trapped?" Or to an EMS call, "What is wrong with the patient?
Is it minor or life-threatening? Is the caller really in distress,*

or is he waiting for us with a 2x4 or a gun?

**I wish you could be in the emergency room as a doctor
pronounces dead the beautiful little 5 year old girl that I have
been trying to save during the past twenty-five minutes who
will never go on her first date or say the words,
"I love you, Mommy" again.**

**I wish you could know the frustration I feel in the cab of the engine,
the driver with his foot pressing down hard on the pedal,
my arm tugging again and again at the air horn chain,
as your feet fail to yield right-of-way at an intersection or in
traffic. When you need us, however, your first comment upon
our arrival will be, "It took you forever to get here!"**

**I wish you could know my thoughts as I help extricate a girl of
teenage years from the mangled remains of her automobile,
"What if this were my sister, my girlfriend, or a friend?
What were her parents' reactions going to be as they opened?
the door to find a police officer hat in hand?"**

**I wish you could know how it feels to walk in the back door
and greet my parents and family, not having the heart to tell
them that I nearly did not come back from the last call.**

**I wish you could feel my hurt as people verbally, and sometimes
physically, abuse us or belittle what I do, or as they express
their attitudes of, "It will never happen to me."**

**I wish you could know the physical, emotional and
mental drain or missed meals, lost sleep and forgone social
activities, in addition to all the tragedy my eyes have viewed.**

**I wish you could know the brotherhood and self-satisfaction
of helping save a life or preserving someone's property, of
being there in times of crisis, or creating order from total chaos.**

*I wish you could understand what it feels like to have a little
boy tugging at your arm and asking, "Is my Mommy okay?"
Not even being able to look in his eyes without tears from your
own and not knowing what to say. Or to have to hold back a
long-time friend who watches his buddy having rescue breathing
done on him as they take him away in the ambulance.
You know all along he did not have his seat belt on, a
sensation that I have become too familiar with.*

*Unless you have lived this kind of life,
you will never truly understand or appreciate who I am,
what we are or what our job really means to us.*

I wish you could...

Author Unknown

Great Fire Disasters and Lessons Learned

The fire service in general, unfortunately takes a reactive approach rather than being proactive. True, we can attempt in vain and "what if" endless possibilities, but usually for something good to happen, sadly, something bad has to occur. Every once in awhile, we need to stop, look, listen, and reflect on who we are and what we stand for, and learn why we do what we do. Many hard lessons learned in the past have resulted from some tragic loss of life and property, but because of it, countless lives have been spared and properties have been saved.

The earliest recorded fire prevention efforts come, once again, from ancient Rome. It was indisputably noted after a major fire, that something should have been done to prevent the disaster or at least, have somebody to protect the citizens. This would also be the grounds for some of the first organized firefighters.

Always on the cutting edge, Rome constantly dabbled with firefighting and prevention efforts. Slaves were used as some of the first firefighters in Rome, and when the slaves who were serving became unruly, the response was to form the equivalent of what we now term a fire department. Once again, this was a reaction to a problem.

The pages of history are plagued with catastrophic stories of great cities that were nearly destroyed by fire:

- **London - 798, 982, 1212, 1666**
- **Venice - 1106, 1577**
- **Boston - 1631, 1653, 1679, 1872**
- **Moscow - 1752**
- **Rome - 1764**
- **Chicago - 1871**
- **Baltimore - 1904**
- **San Francisco - 1906**

After the fact, something significant took place after each of the fires wreaked havoc to the city. Today, the American Fire Services foundation sits upon historic catastrophes that took place. In the wake of each succeeding disaster, improvements and progress were made.

Chicago has become the theater of many historic fires. The city was rebuilt following the "Great Chicago Fire" of 1871. Improvements that came about were made in construction styles and methods. By 1903, the Great Fire had become an evaporating memory to all, as continuing waves of immigrants from the Old World gorged the city. The Iroquois Theater became one of the great entertainment venues in this thriving industrial center. It was one of the big stops on the vaudeville circuit, drawing the great performers of its day. So it was on Dec. 30, 1903, that the stage was set for another epic lesson in fire safety, one that would come at a great loss of lives.

The comedian Eddie Foy was starring in a matinee performance of the musical comedy, "Mr. Bluebeard." A standing-room-only audience estimated at 2,000 people crowded the theater. At some point during the performance, a spotlight overheated and burst into fire way up in the stagehand's overhead. The fire that ensued quickly spread through the flammable backstage rigging. Workers desperately attempted to beat the fire out with sticks to no avail; the blaze was now growing fiercely. A piece of flaming cloth fell to the stage indicating to the guests that something was wrong. In a futile attempt to calm the crowd, Foy had the band continue playing. Suddenly, a woman cried out and the audience made a mad dash for the exits. As the firemen moved in to extinguish the flame, they were met with a tangle of human bodies, all entwined and badly burned.

The cause of death for many came from the terrible smoke and flames. The majority, however, had been trampled and crushed in the rush to leave the hall. Authorities considered it a miracle that only 602 people succumbed to the fire.

The Iroquois Theater was built of fire-resistive materials. Experts agree that it was well built, but they also point out that many important fire protection features were missing or inoperable at the time of the fire. These include:

• Blocked asbestos curtains.
• Installed ventilators that were not in operation.
• Exits not properly marked.
• Exits blocked with draperies, wood and glass doors.
• No installed alarm system.
• No fire protection devices such as extinguishers and standpipes.
• No automatic sprinklers in the stage area, even though it was a municipal requirement.

The investigation that followed led to a variety of fire safety improvements, all of which addressed the problems listed above.

In fact, many cities still require a uniformed firefighter or group of firefighters to stand by at major entertainment events.

The fire service has long viewed old-style factory buildings as a serious fire hazard. Many of the most horrible fires during the late 1800s and early 1900s occurred in factory buildings. Some are more famous than others. A 1910 fire in a Newark, NJ, clothing factory killed 24 workers, and there were countless others. All had code-related violations.

It was the 1911 fire at the Triangle Shirtwaist Company in New York City, which marked a turning point in how fire codes addressed this type of occupancy. Chief Edward Croker of the New York Fire Department had long sought after improvements in the building codes and factory laws, because of such early fires. Three firemen were killed in a massive collapse within this 20-story fireproof building. Sadly, his pleas fell on deaf ears and the resulting catastrophe, which killed scores of innocent young immigrants, will long live in the archives of firefighting teachings.

The Triangle Shirtwaist Company was located at 23 Washington Place in lower Manhattan. The work force was primarily made up of young, female immigrants, who labored under classic "sweatshop" conditions. More than 500 workers were jammed into the eighth and ninth floors of the 10-story building, which was supposedly built from fire-resistive materials.

It was about 4:45 P.M. on Saturday, March 25, 1911 when a fire started in a rag bin on the eighth floor. It spread quickly through the mix of combustible cloth, and soon the contents were ablaze. One group of workers grabbed the standpipe hoseline and without success, attempted to extinguish the fire. They quickly found that the hose was rotted and the valves frozen shut. Word of the fire soon began to pass through the workers packed into the loft building. Workers heaved toward the exits with which they were familiar and there they were met with a wall of fire racing up the stairs. Others moved toward another exit, but were blocked by a locked door. When they were finally able to force it, they found that it opened inward. By this time, there were so many people pushing toward the door that the door was jammed shut; people began piling up at this point.

Very few workers knew that the freight elevator was still working. A number of young girls faced with the prospect of a horrible death by fire chose to leap to their deaths from windows on the eighth and ninth floors. Others managed to make it to the roof, and a small number were able to make their way over ladders to the New York University Law School next door.

Bells in New York fire stations began to sound the alarm, but the problems were many. The streets were littered with bodies, making apparatus placement difficult and ladders could not reach the fire or the roof. Once lines were in position, the fire was quickly extinguished. The death toll was 146 people who either leaped to their deaths or were burned or crushed in the panic. The public was outraged as this fire had proved Croker correct. More was needed than just fire suppression.

After an intense investigation, a number of changes were established. A new bureau of fire prevention was created in the fire department and labor laws were passed outlawing many of the practices, which led to the fire. In the wake of this tragedy, work began on the codes, which eventually led to what we know today as the National Fire Protection Association's Life Safety

Code. The ironic part of this story is that the building remained in use for decades after this tragedy.

Even after tragedy strikes change still takes time. Less than a year after the Triangle fire, another major conflagration struck New York, taking the lives of six men. The Equitable Building was a 10-story structure that was comprised of five individual buildings linked together, that covered the better part of a city block in the financial district. Sadly, the fire began in a wastebasket and spread throughout the building. The fire was spotted by an employee, but instead of taking action, fled in fear causing the fire to gain great headway.

The building was literally riddled with dumbwaiter shafts, elevators and multiple unstopped entrances and passages making it very easy for the fire to spread. As the fire grew, it made its way upward through these shafts. Firefighters quickly moved lines down into the basement and began an aggressive attack, unaware of the fire burning above their heads. After discovering the fire above them, a second alarm was issued. As the fire escalated so did, the number of alarms.

To make matters worse, the weather was not cooperating, unleashing heavy gale force winds, blowing freezing water back onto the firefighters, who were battling the blaze. Firefighters attempting to rescue the building's occupants on the roof just missed being killed themselves when the roof the victims were standing on collapsed, hurling them to their deaths. The debris from this collapse also trapped three men in the basement. Unbeknownst to the fire department, these men had made their way into the basement of the building to rescue millions of dollars in negotiable bonds which, if they had burned, would have created financial chaos for their owners. Only through the heroism of Seneca Larke Jr., a full-blooded Native American, were these men saved. While lying on his stomach over the grate where the men were trapped, under torrents of freezing water and falling rubble, he worked with a hacksaw to cut them free.

The death toll from this disaster included the three civilian workers, one of the basement occupants and two fire department members. One of the major lessons learned from this fire was that the latest method of fireproofing structural members had been proven useless. The lessons from the earlier fires had been ignored. In that era, engineers and architects had specified cast iron as the supporting members for a number of large buildings. To protect them from the weakening effects of fire, they had been encased in hollow tie blocks only proving to be ineffective. In the aftermath of these fires, improved fireproofing of structural members was developed.

Some of the hardest lessons learned regarding fire prevention came by three very distinctive school fires.

• The Lakeview Grammar School in Collinwood, Ohio (176 dead).
• The New London Consolidated School in Texas (294 dead).
• Our Lady of the Angels School in Chicago, Illinois (95 dead).

The causes of these fires were all different in nature. In Ohio, it was a cellar fire of unknown origin that roared up the main stairway of the school, trapping the existing students and killing them. They only knew one way out. The fire department was not trained or equipped to fight a fire in the school.

The victims of the Texas fire were killed in a massive gas explosion. Later investigation indicated that questionable construction, installation, and maintenance processes involving the building's heating system appeared to be the culprit in this disaster.

The fatal fire in the Our Lady of the Angels School began as a small trash fire in the basement. This fire then raced up the main stairway and trapped students in the corridor and in their rooms on the third floor. As a sad footnote, many students were found seated at their desks, heads down, as if praying.

Each of these tragic fires led to improvements that benefit schoolchildren all over North America today:

• Exit drills are mandatory;
• Construction practices are according to code;
• More school inspections in most places;
• Greater emphasis on installed fire protection, alarms, and first-aid firefighting equipment.

Regrettably, many innocent children had to pay the penalty for the sins of adults who did not know or care about fire safety issues.

There are also a number of classic fires in places of public assembly that have led to upgrades in fire and life safety. Some of them are:

• 1903 - Iroquois Theater, described above (602 dead).
• 1919 - Dance Hall, Via Platt, LA (25 dead).
• 1929 - The Glen Motion Picture Theater in Paisley, Scotland (70 dead).
• 1940 - Rhythm Club, Natchez, MS (198 dead).
• 1942 - The Cocoanut Grove, Boston, MA (491 dead).

• 1977 - Beverly Hills Supper Club, Southgate, KY (164 dead).
• 1990 - Happy Land Social Club, Bronx, NY (87 dead).

In each one of these cases, people died in great numbers because fire safety issues were either ignored or never fully addressed.

The Cocoanut Grove was a one-story nightclub that had been built during the Prohibition era. It was a popular site and was constantly loaded with customers. The night of November 28, 1942, was no different. The official occupancy was supposed to be 600, but estimates from that fateful night ranged as high as 1,000.

A small fire started in the basement lounge and quickly raced through the area. Most people knew only the main entrance, which they always used. As the crowd surged toward the exit, it quickly became blocked. Fire department sources listed nearly 200 people as being found in this particular area alone. All told, fire, smoke, heat or the effects of being trampled killed 491 people.

The Boston Fire Department was on the scene quickly for this fire, as a full alarm response had been made to a nearby area for a box alarm that turned out to be a car fire. The fire quickly escalated to five alarms, but the damage had been done. What was learned from this fire?

• Combustible materials must not be used for decorations or in building components.
• Occupancy limit requirements should be strictly enforced.
• Exits need to be kept clear of obstructions and plainly marked.
• Public assembly buildings must have two separate means of egress, remote from each other.

• Exit doors should swing in the direction of egress traffic flow.

The Beverly Hills Supper Club fire in 1977 exposed us to what we thought was a thing of the past. This club was a major regional entertainment center, with many of the country's top entertainers appearing in its lounges. The building was originally erected in 1937. A 1970 fire occurred prior to the building being remodeled. Further expansion occurred in 1974, when the large Cabaret Room was created. It should be pointed out that automatic sprinkler, alarm and kitchen hood fire protection was not installed during any of the construction phases.

Fire struck the club on May 28, 1977. The club was crowded with patrons who hoped to attend one of the John Davidson shows, which were scheduled for 8:30 and 11:30 P.M. At about 8:45 P.M., employees discovered a fire in the Zebra Room. There appears to have been about a 15-minute delay in notifying the fire department while the employees attempted to extinguish the fire themselves.

Many people stated that the first time they noted a problem was when they noticed a large number of people suddenly departing the building. There was no building fire safety plan, so word of the fire was spread from word of mouth. The evacuation appeared to be calm until thick, dark clouds of choking smoke engulfed the exit access areas. One hundred sixty-four people were killed in a tragedy that never should have taken place. The lessons learned in 1942 were unfortunately not remembered, causing history to repeat itself. The list of causes included:

• No installed fire protection.
• No fire safety plan.
• Blocked exits.

• Crowd in excess of the occupancy load.
• Inadequate exit capacity.
• Combustible wall coverings.

In a case study of this incident, Francis L. Brannigan came to the conclusion that the fumes which killed so many of the patrons were not attributable to the electrical wiring. His review of the event, and all documentation, led him to hypothesize that the fumes came from a combustible metal deck roof over the Cabaret Room. It was his finding that heat could have moved up through wiring openings in the walls of the building, which would lead heat up to the decking.

It was his opinion that sufficient temperatures could have been reached that would raise the combustible roof materials to the point at which they would burn. He also stated that even if sprinklers had been installed, the fire would have continued raging above the sprinklers, with the resultant generation of toxic fumes. This event was similar to the giant warehouse fire which occurred during 1985 at Tinker Air Force Base in Oklahoma. He reflects that this could have led to the first large loss-of-life fire in a sprinklered building.

We would venture to say that many of these hard-earned lessons are still being violated. The misnamed Happy Land Social Club had a fire in 1990, which had all of the same issues of crowding, inadequate exit capacity and a lack of installed fire protection. What differed in that situation was an irrational man, hell-bent on revenge, who in the end killed scores of innocent people.

Much of what has been discussed so far has been dealing with serious fires in buildings. There however have been disastrous fires that have occurred in other venues as well. On June 15,

1904, marine catastrophe struck New York City. At 9 A.M., the excursion steamer General Slocum moved away from the 13th Street Pier on the East River jammed with 1,400 men, women and children. It carried most of the Sunday school members from St. Mark's Lutheran Church in Manhattan. They were on their way to a peaceful summer picnic.

Not long after they began their journey up the East River, a fire broke out at the front of the ship. Soon an unforeseen chain of events would be set in motion that would turn a simple boat ride into a death trap. Crewmembers deployed a hoseline to attack the fire, but the hose ruptured as they struggled to turn on the water. As passengers donned life preservers, they quickly found them unusable. Lifeboats were lowered incorrectly, dumping many of the passengers into the water. There was a stiff breeze blowing but that did not stop the ship's captain from turning into the wind in a vain attempt to reach nearby land, which in turn, caused the fire to move quickly through the ship. One thousand twenty-one people were killed on this sad day; whole families were wiped out. After an extensive investigation, a number of substantive changes were made:

• Evacuation drills for crew and passengers.
• Firefighting training for crewmembers.
• Sufficient amounts of the proper firefighting equipment.
• Periodic inspections to ensure equipment condition.
• Sufficient life preservers and lifeboats.

In the wake of the General Slocum disaster, hundreds of ships were inspected for safety-related concerns. Most had the same violations.

Thirty years later, fire on the high seas was once again in the news headlines. The Morro Castle was a popular cruise ship plying the New York-to-Havana tourist run. Unlike the General Slocum, the Morro Castle was protected by a fire detection system. On Sept. 3, 1934, while returning to New York, a fire was detected in a vacant writing-room locker. The ship was equipped with fire doors, and had the door to this area been closed the fire probably would have been held to the room where it started.

While the ship had all of the required fire protection devices, later investigations would show that the crew was not well trained in their use. There were also problems in boarding and lowering the lifeboats. By the time the burned-out hulk of the Morro Castle floated ashore in Asbury Park, NJ, 137 people had lost their lives. Once again, the lessons of the past had been lost on a new generation.

Fire knows no bounds in its drive to kill and maim the innocent. A hot, humid July day greeted the 7,000 people who had chosen to attend the special matinee program of the Ringling Brothers-Barnum & Bailey Circus in Hartford, CT. The extra session had been scheduled due to the late arrival of the circus a day earlier.

The circus owned a large number of portable extinguishers and water buckets, but they had not been placed around the area on July 6, 1944. A major oversight that went undetected by local fire authorities would help encourage the impending doom. While the large main tent had been well cared for, it had been waterproofed in a very dangerous manner. It had been coated with a covering of paraffin that had been thinned by using gasoline making the whole circus tent a highly flammable covering.

As the second act was about to begin, a small flame was observed by an on-duty Hartford policeman. Slowly the fire spread up the tent, gaining speed as it heated the fuel, which lay just ahead. At about the same time, the circus bandleader saw the fire and quickly had the band switch to playing the famous Sousa march, "The Stars and Stripes Forever." This is the traditional circus alarm call.

As the fire grew in intensity, the tent area became a scene of sheer terror and pandemonium. People were pushing toward the main exit, animals were running loose, and the burning tent was falling all around. When the flames were finally extinguished, 168 people lay dead; more than half of which were children.

As a result of this tragic fire, a number of changes were made to improve circus and outdoor event fire safety:

• The National Fire Protection Association (NFPA) formed a technical committee to deal with problems of this nature. The result of this effort led to the development of NFPA Standard 102, Grandstands, Folding and Telescopic Seating, Tents and Membrane Structures.

• Tent tops and tarps must be made from fire-resistive materials.
• Tents are relegated to a temporary role.
• Tents must be properly spaced so that they are not too close to one another.
• Life Safety Code compliance is mandated.

Today fire prevention is stronger than ever with more communities and fire departments embracing the idea. Much of what we see today regarding fire prevention efforts arose from these great tragedies. Thankfully on a daily basis, lives are spared and properties have been saved as results from our past ignorance. Sadly, though, just as the Cocoanut Grove's lessons faded with time, we will as a society become arrogant once again and lives will be lost. Pompous attitudes need to be checked at the door and the value of a human life needs to become immeasurable. Lives have already been lost and lessons have already been learned. Let us remember and not forget.

Firefighter's Prayer

When I am called to duty, God, whenever flames may rage;

Give me strength to save some life, whatever be its age.

Help me embrace a little child before it is too late

Or save an older person from the horror of that fate.

Enable me to be alert and hear the weakest shout,

And quickly and efficiently to put the fire out.

I want to fill my calling to give the best in me,

To guard my every neighbor and protect their property.

And if, according to my fate, I am to lose my life;

Please bless with your protecting hand my children and my wife.

Author unknown

"Firemen are going to be killed right along. They know it, every man of them... firefighting is a hazardous occupation; it is dangerous on the face of it, tackling a burning building. The risks are plain.... Consequently, when a man becomes a fireman, his act of bravery has already been accomplished."

FDNY Chief Edward F. Crocker 1908

Mayday

"Mayday, mayday, mayday!" Radio traffic halts and emotions set in as a call for help from a brother-in-arms echoes through the hearts of all that are listening. As the incident commander attempts to orchestrate a sense of organization out of chaos, a plan of attack is put into motion to rescue him. The eerie soul piercing sound of a personal alert safety device drowns out everything else. The events unfold with confusion and disbelief, the picture becomes all too clear; Some Town, U.S.A will be planning a firefighter's funeral in the morning. A wife will open the door to find representatives from the fire department, a Chaplain, and the words she never wanted to hear. The new widow sits down while fighting back the tears that have already ruined her mascara and tries to answer a question for which there is no right answer. "Mommy, when is daddy coming home?"

Could it happen to me? This question may go through some firefighter's heads from time to time, and maybe not for others. This unforeseen event plays out like a nightmare that we cannot awaken from. Firefighters will die. It might not be today, tomorrow or even next year, but it will happen. It happens in big cities and small towns alike. To think for a second that someone is impervious to such a fateful death makes them naïve and ignorant at the same time.

If one positive lesson can be learned from such tragedies, it might make it right, but for only a minute. We cannot overturn life's hourglass and go back, but we can go forward with knowledge that will make us stronger. No one becomes a firefighter with the idea of an early death, but it happens. Accepting this possible fate and learning from the others before us, it might be enough to hear your grandchildren say, "Grandpa is home!"

A firefighter's funeral is a dark part of the fire services history and traditions, and is not a popular topic of conversation; however it is not one that can be ignored. The following section is dedicated to all that have unselfishly made the supreme sacrifice. They unknowingly have become our silent teachers and will watch us graduate when we attend our own retirement party. As history has the tendency to repeat itself, the list of those answering their last alarms will forever continue to grow.

A Historical Look at Major Firefighter Fatality Fires

It is without question that the profession of a firefighter can be dangerous and even deadly at times. With the increase of fire prevention efforts, added sprinklered buildings, and a more educated public, the amount of fires are drastically down. Rationally thinking, fewer fires would mean less firefighter deaths but this however is not the case. The number of firefighter fatalities have been holding steady with an average of 100 per year since records have been kept. There are numerous factors which contribute to the epidemic of firefighter deaths, such as less experienced firefighters due to the fact that fires are down. One dynamic that hasn't changed is building collapse. In the following section a list of major firefighter fatalities is reviewed dating from the early 1900's to September 11, 2001. Thankfully most firefighter deaths do not come in great numbers as the ones being discussed, so let us not forget the many fires that claim only one or two lives. The information that has been gathered is limited in nature and there may have been some that have not been properly reported or documented. Therefore some of the information may be conflicting.

- **December 21, 1910 Philadelphia, Pennsylvania: 13 firefighters killed** and one police officer in a leather remnants factory.

- **December 22, 1910 Chicago, Illinois: 21 firefighters killed** at a stockyard and cold storage warehouse in Chicago.

- **March 10, 1946 Brocton, Massachusetts: 13 firefighters killed** in a roof collapse at the Strand Theatre.

- **April 16-17, 1947 Texas City, Texas: 27 firefighters killed** following fires and ammonium nitrate explosions aboard two ships docked.

- **August 4, 1949 Montana: 13 firefighters killed** in a forest fire at Gates of the Mountain.

- **July 9, 1953 California: 15 firefighters killed** in a wildfire in the Mendocino National Forest.

- **July 29, 1956 Sun Ray, Texas: 19 firefighters killed** at the Shamrock Oil and Gas Corp. refinery.

- **January 28, 1961 Chicago, Illinois: 9 Firefighters killed** as a result of a wall collapse.

- **October 26, 1962 New York City: 6 Firefighters killed** as a result of a collapse at structure fire.

- **October 1, 1964 Boston, Massachusetts: 5 Firefighters killed** as a result of a wall collapse.

- **October 17, 1966 New York City: 12 Firefighters killed** as a result of upper floors collapsing.

- **October 16, 1967 Cliffside Park, New Jersey: 5 Firefighters killed** as a result of a roof collapse.

- **June 17, 1972: Boston, Massachusetts (Vendome Hotel): 9 Firefighters killed** as a result of a wall collapse.

- **July 5, 1973 Kingman, Arizona: 12 firefighters killed** at the Doxol Gas Company.

- **August 17, 1975 Philadelphia, Pennsylvania: 6 Firefighters killed** as a result of an explosion at the Gulf Oil Refinery.

- **August 2, 1978 New York City: 6 Firefighters killed** as a result of a roof collapse.

- **December 27, 1983 Buffalo, New York: 5 Firefighters killed** as a result of an explosion from a propane leak in a three-story radiator warehouse.

- **July 23, 1984 Romeoville, Illinois: 10 Industrial Fire Brigade Members killed** as a result of an explosion from a propane gas leak.

- **July 1, 1988 Hackensack, New Jersey: 5 Firefighters killed** as a result of an arch-truss roof collapse.

- **November 29, 1988 Kansas City, Missouri: 6 Firefighters killed** as a result of an explosion involving a semi-trailer that was loaded with explosives.

- **July 6, 1994 Glenwood Springs, Colorado: 14 wildland Firefighters killed** as a result of 70 mph winds accelerating the fire spread that trapped them.

- **February 12, 1998 Chicago Illinois: 2 Firefighters killed** as a result of a roof collapse in a car dealership.

- **December. 4, 1999 Worcester, Massachusetts: 6 Firefighters killed** as a result of becoming lost and cut off by heavy fire during a search and rescue operation.

- **December 22, 1999 Keokuk, Iowa: 3 Firefighters killed** as a result of a flashover while performing a search for three trapped children.

- **June 17, 2001 Queens, New York (Fathers Day Fire) 3 firefighters killed** as a result of a wall collapse in hardware store fire.

- **September 11, 2001 New York City: 343 Firefighters murdered** as a result of the twin towers collapsing in the aftermath of terrorist attacks at the World Trade Center.

These firefighters lost their lives while performing the largest search and rescue operation in the history of the fire service.

Bagpipes at Fire Department Funerals

Bagpipes in the fire service have become part of its rich tradition just as baseball and apple pie have become an American institution. How they became one and the same is a different question. The following tale is one that you may have heard before of how bagpipes were introduced into the fire service. While some discrepancies arise from its origins, some believe bagpipes came from the Scottish while others assume the Irish hold the rights, it's roots can be traced back to the Middle East several centuries before the birth of Christ. For certain, bagpipes have internationally become synonymous with acts of bravery and sacrifice.

The tradition of bagpipes being played at fire department funerals in the United States goes back over one hundred and fifty years. When the Irish and Scottish immigrated to this country, they brought many of their traditions with them. One of these was the bagpipe, often played at Celtic weddings, funerals and dances.

It wasn't until the great potato famine in 1845, which triggered a massive Irish immigration to the East Coast of the United States, that the tradition of the pipes really became predominant in the fire service. During this time jobs were scarce and times were hard. The best jobs they could get were jobs that were dirty, dangerous or both, jobs being that of firefighters and police officers. It was not an unusual event to have several firefighters killed at a working fire. The Irish firefighter's funerals were distinctive of all Irish funerals in that the bagpipes were played. It was somehow okay for a hardened firefighter to cry at the sound of pipes when his dignity would not let him weep for a fallen friend.

Those who have been unfortunate enough to attended a funeral when bagpipes play, knows how haunting and somber the sound of the pipes can be. Before too long, families and friends of non-Irish firefighters began asking for the piper to play for these fallen heroes. The pipes add a unique air and distinction to the somber occasion.

Today, the practice is universal and not just for the Irish or Scottish. The pipes have come to be a distinguishing feature of a fallen firefighter's funeral.

Though some minor accuracy comes from the above story, it dives right in at the shallow end and does not let us swim deeper to the true birth of this fire service ritual. If you haven't had the opportunity to have your soul pierced by the sound of a piper playing at a firefighter's funeral, consider yourself blessed and God willing I wish this not on my worst enemy. Hopefully this will be as close as you get:

Two pipers stand on a hill away from the assembled gathering. The cold wind whips the grasses about their legs as they stare at rows of mourners. The pipers fill their instruments with air and place the bags beneath their arms. With effort and concentration they fill the burial grounds with the drones and wails of the pipes. The music enhances the grief of the family of the dead and the other mourners. It also serves to escort the fallen to the final resting place.

This scenario has been recreated for centuries with the tradition being carried onward from the ancient battlefields of Ireland and Scotland to the ceremonies honoring fallen peace officers and firefighters. Contrary to popular belief, the bagpipes are not of Scottish or Irish decent. The first account of the instrument can be traced back to the Middle East several centuries before the birth of Christ. It was most likely a rather crude instrument consisting of reeds stuck into a goatskin bag. As civilization spread throughout the Middle East and into the Mediterranean lands, the people brought along their music.

It is commonly believed that village musicians used the pipes along with drums for entertainment in an effort to brighten an otherwise dull existence. The instruments also contributed to other facets of community life, such as the burying of the dead. Some of their instruments were variations of the early bagpipe. Instrument complexity grew with time, as reeds were replaced with the hollowed leg bones of small grazing animals. Holes drilled into the bones allowed the musicians to vary tones and pitch.

Alongside their trumpets and drums, the Romans carried with them their pipes. The legions spread the popularity of the instrument throughout Europe, eventually ending up in the British Isles. It was in the majestic settings of Scotland and Ireland that the pipes reached their highest level of recognition. During the 14th century, the bagpipes could be found in nearly every village.

In addition to providing music for enjoyment, bagpipes were also used to unite the clan to battle, usually against the English. The English found the pipes so unsettling that they banned the Scots and the Irish from playing them at any time. Yet, that didn't stop their use. Over the years, the bagpipes grew in sophistication with more pipes being added, enabling the musician to reach a wider range of notes.

Somber melodies were often played over the pipes following the death of a fighter. Villagers would congregate around the body of their dead compatriot while the piper stood a distance away. It was said that the music was a sign to God that a courageous and devoted subject was on his way to sit at "the steps of the throne."

The bagpipe was launched into the new world during the great migration westward. British soldiers were led into battle by pipers and drummers. As the battle began, the musicians peeled off to the side, encouraging the soldiers to battle on. Songs of victories drifted over the battlefield, reminding the soldiers of their heroic legacies.

During the great Irish potato famine of the 1840's, hundreds of thousands of sons and daughters of Ireland left their homes for the promise of a better life in America. During the dreadful journey across the Northern Atlantic, the only interval was their music. Bagpipes, flutes and drums brought back memories of their heritage, and when they could, they sang and danced to the tunes. Their music also grieved for fellow travelers who fell sick and died along the way.

With a fervor possessed of many immigrants, they labored their way into the mainstream of America taking jobs of firefighters and police officers; the most dangerous occupations at the time. They also joined the armies of America during the Civil War, entering regiments that were made up of Irish descendants and immigrants, and the bagpipes and drums were again brought to the fore to lead men into battle.

Tolling of the Bell

The crackling speakers of the stations alerting system break the stillness of the night. Sleeping firefighters are awakened from their light slumber, because no one really ever gets a good night sleep while at the firehouse. Heart rates jump from a peaceful resting rate of 40-50 beats per minute to an astounding 150-180 beats per minute in just a matter of seconds. Going from 0-60 is an all too familiar episode for firefighters around the country. In order to alert firefighters of a call, there are now many different methods used ranging from traditional bells or gongs, to modern paging systems and "tones." Whichever system is used, all can be traced back to earlier colonial times.

Firefighting in early American times goes all the way back to Jamestown, Virginia during the days of the "Rattle Watch". This was generally documented as the first way to alert firefighters in America. Rattles soon gave way to bells or other means of warning citizens of a community of a fire.

As the fire service developed so did its methods of alerting firefighters. Long before the internet was invented, or telephones and radios were available, fire departments used the telegraph to communicate using special codes to receive fire alarms from those once familiar red fire alarm boxes which stood on practically every street corner of America. These special codes were also used to communicate between stations sending important messages.

In the past and some presently, as firefighters began their tour of duty, it was the bell that signaled the beginning of that days shift. Through the day and night, each alarm was sounded by a bell, which summoned these brave souls to respond to emergency after emergency. When the fire was out and the alarm had come to an end, it was the bell that signaled to all, the completion of that call and that the company has returned to quarters. The signal was three rings of three (3-3-3).

Different cities had different interpretations of this "return to quarters signal." The tolling of the bell, had different cadences from three rings of three (3-3-3), five rings of three (5-5-5) to five rings of four (5-5-5-5). Which ever was used, this recently has become the signal for a firefighter killed in the line of duty. When a firefighter's life was taken, or in the language of the military and public safety: "fell", in the line of duty, the fire alarm office would tap out this special signal.

This came to be called the "Tolling of the Bell" and was broadcast over the telegraph fire alarm circuits to all station houses in the district. Heard outside on the streets with the firehouse windows open, the resonating echo was similar to that of fire stations of yesteryear, where fire alarm bells sounded the locations of thousands of emergencies throughout the history of our growing country.

During times like these we seek strong symbols to give us a better understanding of our feelings during this time of sadness and as a reflection of the devotion our comrade had for his duty. Tradition carries on at a firefighter's funeral. The reading of the "Firefighter's Prayer" and the presentation of his battered helmet to the family are all traditions which gives honor and respect to those who have served so well. So, also, is the sounding of a bell. The tolling of the bell represents the end of his duties and that he will be returning to quarters. For our comrade this is his last alarm. He is coming home.

The Last Alarm

My father was a fireman, he drove a big red truck

And when he'd go to work each day he'd say

"Mother, wish me luck"

Then dad would not come home again until

sometime the next day

But the thing that bothered me the most was the

things some folks would say

A fireman's life is easy, he eats and sleeps and

plays,

and sometimes he won't fight a fire for many, many

days

When I first heard these words I was young to

understand

But I knew when people had trouble Dad was there

to lend a hand

Then my father went to work one day and kissed

us all goodbye

but little did we realize that night we all would cry

My father lost his life that night when the floor

gave way below

And I wondered why he'd risked his life for

someone he did not know

But now I truly realize the greatest gift a man can

give

is to lay his life upon the line so that someone else

might live

so as we go from day to day and pray to God

above

Say a prayer for your local fireman

He may save the ones you love.

<u>Emergency Medical Services</u>

Paramedics Prayer

LORD
Grant me the wisdom so that I may treat
those of your children that lay at my feet.
Let my hands be gentle, sure and swift
to impart to them your sacred gift.

Let me see only a patient's need
not their color, race or creed.
Help me always to be my best
even when it's on my hours rest.

Grant me the insight to understand why
patients of mine are going to die.
Let me remember that when they do
there is a wonderful life in Heaven with You.

Lord, if in the time of duty I should fall
help my family to hold their heads tall.
For it was You who decided that I should be
one of your chosen few, an EMT.

-Author Unknown-

Star of Life

In the year 1966, much was going on. The Viet Nam War, Lyndon Johnson was in office, and Emergency Medical Services (EMS) was in its early stages in the United States. The Star of Life would soon be an easily recognized symbol of EMS. It now is seen on ambulances, signs directing people to emergency rooms, and on paramedic patches. Unfortunately, most people, including many EMS people, don't know the significance or history of the Star of Life.

Prior to 1966, it was a familiar practice for patients to be transported to the hospital in vehicles operated by funeral homes or in some areas of the country by the police department. The hospitals that the patients were transported to had very few doctors properly trained in emergency medical care. In the late 1960's, a study by the National Academy of Sciences was published called *Accidental Death and Disability: The Neglected Disease of Modern Society*. The paper reviewed motor vehicle collisions and the injuries they caused. However, it also studied the "system," or lack of an emergency medical care system, that was in place at the time. It acknowledged many insufficiencies in emergency care and it was time to make a change in America and develop a suitable emergency medical care system.

The Department of Transportation (DOT) was given the mission of managing a program that sanctioned planning for ambulances and the equipment that would go in them, after Public Law 89-56 was approved in 1966. Under the Department of Transportation, Emergency Medical Technician (EMT) standards were perfected, paramedic standards were created, and ambulances were purchased for EMT and paramedics to use when responding to calls. To help identify these emergency care personnel and vehicles, a symbol needed to be designed. It would become known as the "Star of Life."

In the beginning, EMS used a red cross with four bars, which is a trademark of the American Red Cross (ARC) and the International Red Cross (IRC). EMS organizations in the late '60's and early '70's used it or an orange version called "Omaha Orange". The ARC did not like EMS using the cross and stated it was a violation of an international treaty, backed up by federal law. The treaty and law states that the red cross in any form cannot be used by anyone not connected with the IRC. Johnson & Johnson is the only corporation that can do this because it was grandfathered under the federal law.

Since the Red Cross emblem could not be used, Leo R. Schwartz, then Chief of the EMS Branch, NHTSA, took the red Medical Identification Symbol (Medic Alert) of the American Medical Association, changed it to blue, and placed it on a white square. This new Star of Life had six bars with a serpent entwined around a staff in the middle; this has become the symbol of emergency medical services.

The following is the story the surrounds the rationalization for the serpent and staff representation.

The snake and the staff in the center of the Star of Life portray the staff of Asclepius who, according to Greek mythology, was the son of Apollo god of light, truth, and prophecy and the mortal maiden Coronis. Apollo was told to instruct his son in the ways of medicine and healing. Conflicting chronicles report that Asclepius either learned the art of healing from Apollo or Cheron, the centaur. Regardless of who taught him, Zeus, king of the gods, was fearful that with Asclepius' knowledge, men might be rendered immortal. Rather than have this occur, Zeus killed Asclepius with a thunderbolt.

After Asclepius' death, he was worshipped and thought to be a god of healing. People would sleep in his temples. He would appear to them in their dreams offering cures and remedies. Drawings of Asclepius usually show him in a standing position, dressed in a long cloak, holding a staff with a serpent coiled around it.

Another reason for using the serpent and staff may come from the Bible, in Numbers 21:8 and 9 which makes reference to a serpent on a staff.

And the Lord said unto Moses, Make thee a fiery serpent, and set it upon a pole: and it shall come to pass, that every one that is bitten, when he looketh upon it, shall live. And Moses made a serpent of brass, and put it upon a pole, and it came to pass, that if a serpent had bitten any man, when he beheld the serpent of brass, he lived. (Bible 196).

The six bars of the Star of Life represent six distinct phases of an EMS response which are **detection, reporting, response, on scene care, care in transit**, and **transfer to definitive care**. These phases are considered critical to producing an excellent outcome for the patient. The phases are:

1. **Detection:** Citizens must first recognize that an emergency exists and must know how to contact the EMS system in their community. This can be by several different methods such as dialing 9-1-1, using a seven digit local emergency number, or using amateur radios, or highway call boxes.

2. **Reporting:** Callers are asked specific information so that the proper resources can respond. In an ideal system, certified Emergency Medical Dispatchers (EMDs) ask a pre-defined set of questions. If someone were having a heart attack, then they would look under the heart attack algorithm for appropriate questions to ask and also give appropriate pre-arrival instructions (such as CPR). In this phase, dispatchers also become a vital link between the scene and the responding units and can provide additional information as it becomes available.

3. **Response:** This is the response of the EMS resources to the scene. This may be a tiered response with First Responders and EMTs responding initially and backed up by paramedics shortly thereafter. It may mean that a fire engine and crew are also dispatched to assist with lifting and moving the patient, patient care or assisting with a motor vehicle crash.

4. **On Scene Care:** Various types of care can be rendered on the scene, versus waiting until the patient arrives at the hospital. Standing orders and radio or cellular contact with the emergency physician has broadened the range of on-scene care that can be provided. A long algorithm of procedures and drugs may be used before the patient is removed from the scene.

5. **Care in Transit:** As stated earlier, patients were once transported in hearses or station wagons, with nobody taking care of them in the back. With the advent of federal regulations and the maturing of EMS, ambulances now carry mobile oxygen, suction, patient monitoring and communications equipment, as well as special drugs for emergency care of patients.

6. **Transfer to Definitive Care:** Up until the passage of the Trauma Care Systems Planning and Development Act of 1990, a patient might be seen in the emergency room (ER) by a physician trained in a certain kind of specialty, such as a cardiologist or a surgeon. In the past they did not have the training necessary to address the many types of injuries and illnesses that present themselves in an ER. Today, there are board certified emergency medical care physicians waiting to help patients. Nurses now receive certification in emergency care and specialized training in trauma. Hospitals may hold special levels of designation in trauma care. This means they have additional specific equipment, rooms and physicians available for the most traumatically injured patients. There are specialized burn centers to handle burn patients and special children's hospitals that handle only pediatric patients. Definitive care has come a long way, as has EMS, in a relatively short time.

Originally, the Star of Life was supposed to be something different for the different types of EMS personnel. For administrative and dispatch personnel, the Star of Life was to have a silver colored edge and the staff of Asclepius, a silver colored serpent. For EMT's and Paramedic's patches, the edge was supposed to be gold in color with a gold serpent. There is also a variation (a white star on a blue background) that was going to be used to categorize receiving facilities, such as hospitals or emergent care facilities

EMS is a system and the Star of Life represents that system. The system has matured vastly over the past thirty years and will continually be changing and evolving rapidly during the 21st century. The continued development of EMS depends on an educated public being able to recognize when to call for emergency help, the availability of specially trained people and equipment, and a network of specialized trauma centers. The Star of Life symbolizes these elements and, hopefully, once people understand the meaning of the Star, they will understand what actually makes up the EMS system.

History of the Ambulance

From earliest times, people have required a means of transporting their wounded and sick to a place where they could receive definitive care. The very first designs, which are very primitive by today's standards, laid the groundwork for what we know today as the modern ambulance. Just as patient care has blossomed so did the evolution of the ambulance. We will take a walk through time and explore this slow process and examine just how the concept possibly started with a hammock strung between two poles.

It is thought that quite possibly the first form of patient transport may have been as simple as a hammock strung between two poles. Even though this is a far cry from what we view today as an ambulance, it may have been the beginning. Aside from the hammock and pole theory, other unique conveyances included human dhooleys that were used in India in which wicker cradles were called mule panniers. In Egypt, a similar model was used, but with a different name. There they were called camel litters. The Mojave Indians also devised a litter comprised of a sheet of canvas between two poles carried on the shoulders of two men.

A more formalized ambulance surfaced in the late 15th century, when Ferdinand and Isabella of Spain took an unprecedented interest in the welfare of their troops during their crusade against the Moors. Surgical and medical supplies were brought together in special tents for the wounded called ambulancias.

In the late 1700s, Napoleon Bonaparte appointed Baron Dominique-Jean Larrey to develop the medical patient care system for the French army. One of findings was that leaving wounded soldiers on the field for several days increased the complications and suffering. He felt that this delay in treatment resulted in needless deaths. "The remoteness of our ambulances deprived the wounded of the requisite attention," he wrote. In 1797, Larrey developed a method to send trained medical personnel onto the field to provide medical care to the wounded soldiers and to provide medical care en route to the field hospital. This action increased their chances of survival and benefited Napoleon's conquest efforts. He designed a special carriage staffed with medical personnel to access all parts of the battlefield. The carriage became known as the ambulance volante, or flying ambulance. This idea he later refined for Napoleon during the General's Italian campaign.

The evolution of the ambulance took yet another turn during the American Civil War when ambulances were too few, often late, and driven by civilian drunkards and thieves. A physician named Jonathan Letterman reorganized the field medical service to provide an effective ambulance service for the evacuation of battle casualties. In 1864, an act was passed in Congress entitled "An Act to Establish a Uniform System of Ambulances in the Armies of the United States," which spelled out who was responsible for each phase of the ambulance system.

During the 1864 Convention in Geneva, an agreement was made by several European countries to recognize the neutrality of hospitals, of the sick and wounded, of all persons connected with relief service, and the adoption of a protective sign or badge. In America, a similar organization had been functioning during the Civil War. The Sanitary Commission, which 20 years later became the American Red Cross, was brought into being in large part due to the efforts of a woman by the name of Clara Barton.

Most ambulance innovations took place during wartime, which were then adapted to civilian life. American hospitals initiated their own ambulance services during the late 1860s. Horse drawn, these ambulances had a moveable floor that could be drawn out to receive the patient. Beneath the driver's seat was a container with: a quart of brandy, two tourniquets, six bandages, six small sponges, splint material, blankets and a two-ounce vial of persulphate of iron. With the arrival of the automobile came a different type of ambulance first appearing in 1899. During World War I, many ambulances were adapted from buses and taxis. In 1867 Major General Rucker won the "best of kind" for an ambulance that was adopted as the regulation ambulance. It had extra springs on the floor, more elasticity to the stretchers, and improved ventilation.

The world's oldest builder of ambulances is the Hess and Eisenhardt Company in Cincinnati, Ohio. In 1937, they sold the first air-conditioned ambulance built in America. Developed with the idea that the ambulance should be a pre-hospital emergency room, these pre-cursors of the modern ambulance were filled with medicine cabinets, roof lights, and two-way radios.

The first ambulance service in the United States was created in Cincinnati in 1865 at Cincinnati General Hospital. This service is still operated by the fire department. Other services followed at Grady Hospital in Atlanta, Charity Hospital in New Orleans, and several hospitals in New York City and other major cities. In December of 1869 the first month of operation of the ambulance service of the Free Hospital of New York (Bellevue) ran 74 calls. A total of 1466 calls were run in 1870. The dispatch system was much different from that used today. The hospital rang a bell, which triggered a weight to fall, lighting the gas lamp to wake the physician and the driver. It also caused the harness, saddle, and collar to drop on the horse and opened the stable doors; however, this improved care was mostly limited to the larger cities.

Ambulance service has not been confined to ground units. During the Civil War, train ambulances and steam boat hospitals were used, and street car/trolley ambulances were popular in some cities in the late 1800s. More recently, the Star of Life, a water cruiser, began to operate under the direction of the Coast Guard and Marine Police in 1976.

The "medicopter" is now commonplace. It first came into service as a means of evacuating combat patients in Korea and Vietnam, and is now invaluable during natural disasters, such as the Johnstown Flood, when medical supplies required refrigeration and patients needed to be airlifted to hospitals.

Today's ambulances are now literally emergency rooms on wheels and come equipped with some amazing new technology. Life saving equipment such as defibrillators and monitors that can transmit a complete 12-lead ECG directly to the emergency department, advanced airways, drugs and intravenous therapy to the latest in handheld computers are just some of the innovative equipment carried. The advancement of the ambulance has definitely come a long way from early battlefield times. As technology advances we will see change in the equipment that is carried, but the framework that is recognizable today will be here to stay for times to come.

The Origins of Paramedics

Compared to the fire service's age, paramedics and the history of Emergency Medical System are still in its infancy. Organization of the paramedic program that continues to be formed today can be traced back to early battlefields when the recognition of rapid transport to a field hospital was the difference between life and death for a wounded soldier. EMS and the level of training that paramedics now receive are leaps and bounds from the early "load and go" days. With the rise in emergency medical calls ranging from trauma, medical, and cardiac among others, and with fires being down, there will always be a need for advanced emergency medical care. Exploring the history of paramedics and the initial stages of EMS needs to reviewed so as to not lose what will someday be thought of as primitive compared to where it will be one day in the future.

"Paramedics"

The word "Paramedic" is one with origins that are inexact. Even though the original use of the word in a medical context is lost to undocumented history, it is believed by some that it developed out of military usage, referring to those in the military who are medically trained and are paratroopers. According to certain dictionaries, the prefix, "para," is defined as "beside; beyond; against; apart from" and the word "Medic" as slang for "Doctor" or, in Army terms, a "Medical aid man."

As early as 1967, in the United States, the expression "paramedic" reached accepted use in relating to non-physicians (in fact, usually firefighters) who were trained to perform certain advanced life support functions in out-of-hospital settings. In

1970, these vaguely portrayed individuals were first defined by law, and list by California's Wedworth-Townsend Paramedic Act, who listed them, as "Mobile Intensive Care Paramedics."

Inside five years after the California law was approved, 29 other states had established different identities by law. No less than 15 official titles were recognized in statutes to explain the new and evolving variety of allied health care technicians who took their emergency care skills to the streets. The titles ranged from "Cardiac EMT" and "Cardiac Rescue Technician" to "EMT-Advanced," "Advanced EMT" and "Physician's Trained Mobile Intensive Care Paramedic."

Soon, two federal agencies, the U.S. Department of Health, Education and Welfare (DHEW) and the U.S. Department of Transportation (DOT) along with the funds provided by these agencies would play a major role in the development and enhancement of emergency medical services (EMS) in upcoming years. Thus, when these agencies decided to use the term "paramedic," the debate over titles was all but over.

During February of 1975, the DHEW published its Program Guidelines, which included a glossary of terms. This glossary defined a group of "persons trained for advanced life support services to include sophisticated trauma, cardiac care, and other critical care elements for intravenous therapy, shock therapy, drug administration, and cardiac rhythm detection control. The federal document referred to these persons as "EMT Paramedic." Therefore, a middle ground was reached and "EMT" (meaning Emergency Medical Technician) was combined with a word that had irreversibly fixed itself to a common usage in the American vocabulary.

In 1975, the DOT granted a contract to the University of Pittsburgh for development of a standard training program to ready personnel for their role in a pre-hospital setting. The name elected for this course was "National Training Course for Emergency Medical Technician--Paramedic."

Starting in 1971, television entertainment has included a weekly series featuring the dramatized activities of Los Angeles paramedics. The influence of "Emergency!" and later on in the 80's "Rescue 911" has been beyond measure. Since its beginnings, the key figures depicted on "Emergency!" have been identified as "paramedics." In the meantime, millions of American children have been heard to say that they want to be a "paramedic" when they grow up. With that, the word is here to stay.

Closing

What firefighters do on a daily basis is the heart and soul of it all. As you can now see, the history and traditions of the fire service is the mortar that holds together each brick that makes up all firefighters past, present and future. New traditions are always in the make; after all they have to start somewhere. History is the past, but is it also the future. On September 11th, 2001 all who witnessed that dark day were watching history unfold in its ugliest form. That day affected everyone on many different levels. That day 343 firefighters lost their lives. Yet that day inspired many people who may never have given the idea a second thought before, but now wanted to be a firefighter. Each firefighter has their reasons for choosing this calling and one should not judge why. Volunteers, paid on call, part time, full time, men, women, creed and colors of every shade should be judged the same. Everyone in the fire service is an equal and should not be judged on the amount of pay or lack there of, black or white, man or woman.

Many who choose to take on the challenges often go unrecognized for their efforts and often do not want to be. You will over and over again hear a firefighter or paramedic say that they were just there, that any of their brothers or sisters would have done the same thing. This couldn't be any truer. Despite the fact that there are a few that do it for the "glory", I would like to think even the average Joe that just took the job for the pay, does it because they deep down love it. The following is a story of the kid who always wanted to be a firefighter.

As he sits near the television wearing his Spiderman pajamas, he occasionally glances at the clock to see how close it is before mom sends him off to bed.   The front door opens and his father comes and says, "Get dressed. I have something to show you." Confused by the statement and wondering if it's a joke because after all, it's almost bedtime and he has school tomorrow he looks at his mom for her approval. "Get dressed. Put on some boots and a coat, we will not be gone long", his fathers says with some excitement in his voice.

The young boy scrambles to find his "moon boots", you know the snow boots that were popular but were made of cheap material that offered no ankle support and very little warmth. He grabs his coat as his mom says, "Don't keep him out long, he has school tomorrow and don't forget your hat."

Out the door to dad's truck nicknamed Ole' Blue because you guessed it, it was blue. The air was crisp and the wind was calm but temperatures were actually normal for a January night. "Where are we going dad?" the boy asks with curiosity. "You'll see", is the only response he gets.

Only a few seconds later and less than a mile from home, Ole' Blue turns up the snowy drive and instantly the sky becomes an eerie shade of orange. Smoke dances across the night's sky as the silhouettes of firefighters drift in and out of sight from on top of the roof. The scene plays out and what was viewed through that young boys eyes would be a life changing experience. The boy looks at his father and says, "Daddy, I want to be a fireman."

Timeline

1607 --
• First settlement in the "new world" founded in Jamestown.

1609 --
• January 7. Jamestown settlement destroyed by fire. All provisions lost and many die of hunger and exposure.

1613 --
• Dutch trading ship Tiger burns in New York Harbor, forcing her crew to be first settlers on Manhattan.

1620 --
• Pilgrims land at Plymouth.

1623 --
• November 1. Fire destroys seven dwellings at Plymouth and nearly ends settlement.

1630 --
• Boston, MA was settled.

1648 --
• Peter Stuyvesant, Governor of New Amsterdam, appoints four fire wardens to perform inspections of chimneys. Fire prevention ordinances are passed and fines are imposed to purchase and maintain fire buckets, hooks and ladders.

1653 --
• January 14. First of Boston's great fires destroys one-third of the town.

1654 --
• Joseph Jynks of Saugus, MA, builds first American fire "enjyne".

1664 --
• British take over New Amsterdam and the town is renamed New York.

1666 --
• September 2. Great Fire of London England.

1676 --
• November 27. Fire at Boston destroys large part of North End, including Increase Mather's Church.

1679 --
• Boston imports fire engines from England.

1682 --
• Town of Philadelphia is founded.

1711 --
• "Towne House Fire" in Boston.

1718 --
• Mutual Fire Society organized in Boston by progressive citizens. This first volunteer fire company in America only battled fire amongst its member's homes.

1731 --
• Newsham engines arrive in New York from London.

1736 --
• Benjamin Franklin establishes the Union Fire Company, the first fire organization in Philadelphia and one of the first in America to fight fires for the public good of all.

1740 --
• November 18. Fire in Charleston, SC. All houses and buildings from Broad and Church Streets to East Battery are burned down.

1741 --
• March 18. Fort George on lower end of Manhattan burns. New Yorkers incited to rise against "Negro Plot."

1743 --
• First successful pumping engine built in America, by Thomas Lote, New York.

1752 --
• Philadelphia Contributionship started; first successful fire-insurance company in America.

1770 --
• March 5, the Boston Massacre. British troops opened fire on a mob of colonists. The shooting may have started when a wayward cry of "fire" was given -- a cry intended to sound the town's church and fire bells as an alarm.

1775 --
• April 19. Paul Revere's famous ride.
• June 17. Battle of Bunker Hill. Redcoats set fire to 380 dwellings in Charlestown, MA.

1776 --
• July 4. Declaration of Independence.
• September 21. New York burned. American patriots believed to have "scorched earth" for the British by fire that destroyed 493 buildings.

1783 --
• America wins her independence.

1785 --
• A powerful new, American-designed, "Gooseneck" style fire engine was introduced.

1788 --
• March 21. New Orleans destroyed by fire on Good Friday. The fire was started by a curtain blowing over an open candle. Over 900 buildings (seven-eighths of town) burned.

1789 --
• George Washington becomes first President.

1791 --
• Journals recording fire-company duty and activities first used.
• First double-deck/end stroke hand engine built in Philadelphia.

1792 --
• Insurance Company of North America was formed.

1796 --
• December 9. "Coffee House Slip" Fire in New York.

1797 --
• Newark, New Jersey organized a fire company and purchased small hand-engine.

1800 --
• English inventor John Carry, designed the first crude automatic sprinkler but it went undeveloped for a long period.

1803 --
• Fort Dearborn (Chicago) founded.
• Volunteer fire department established in Cincinnati, OH.
• Wooden hydrants installed in Philadelphia, PA.

1805 --
• June 11. Detroit, MI destroyed by fire.

1806 --
• In the aftermath of a major fire in Philadelphia, hydrant inspections began.

1807 --
• Early fire prevention activity in Washington, DC resulted in all chimneys being cleaned.

1808 --
• Fire Bucket Company formed in Cincinnati OH.
• NYC got first hydrant attached to underground water main.
• A new concept in fire fighting, leather fire hose seamed by copper rivets.

1809 --

• The first fireboat (Engine 42 of New York) was hand-rowed and hand-pumped.

1811 --

• Philadelphia diverted water from storage trunks in the street for use in fighting.
• Philadelphia Hose Company formed.
• May 31. Fire destroys 250 buildings in Newburyport, MA

1812 to 1814 --

• The war with England brought major conflagrations: Capitol Building, State and Department, Treasury Building, Government Arsenal all were destroyed by fire.
• "Fire rafts" [burning ships set adrift) were used to threaten and destroy the enemy.

1816 --

• Richmond, VA organized the Richmond Fire Society for the purpose rendering mutual aid.

1819 --

• Philadelphia organized a special fire patrol group to perform salvage work.

1820 --

• January 11. Fire destroys 463 buildings, mostly dwellings in Savannah, GA.

1821 --

• A volunteer fire company is organized in St. Louis, MO.

1827 --
• English chemist John Walter invented the friction match containing phosphorus sulfate.

1825 --
• Erie Canal opened.

1829 --
• George Braithwaite built first fire engine using steam to pump water.

1830 --
• B & 0 Railroad, (the first in America), makes successful run.
• Fire protection improvements in Richmond, VA brought a water supply system reservoir, water pump hose and private hydrants.

1832 --
• Horses began pulling engines after Asiatic cholera plague struck firemen.
• Sectional ladders used by John Braidwood of London.

1835 --
• December 16. Great Fire in New York City. Over 650 buildings, including most of the Wall Street financial area are destroyed and results in $20-40 million property loss. Following the conflagration, private patrols were appointed to cover the city in event of second fire.

1837 --
• Depression and panic sweep America.
• June 10. Broad Street Riot in Boston. Firemen fight Irishmen all day.
• Milwaukee, WI formed its first volunteer fire company.

1840 --
• Henry R. Worthington invented an independent, direct-acting steam pump.

1841 --
• Paul Hodge builds the first steam fire engine in America. He is scorned by the volunteer firefighters of New York.

1844 --
• Samuel F. B. Morse invents the telegraph.

1845 --
• Dr. William F. Channing of Boston invents the fire-alarm telegraph.

1845 --
• April 10. Great Fire of Pittsburgh; 982 buildings burned, mostly dwellings.

1845 --
• Potato famine in Ireland. Irish come to America in great numbers.

1846 --
• July 13. Fire in Nantucket; 300 buildings destroyed. Brought an end to the town's whaling supremacy.

1848 --
• August 17. Fire at Albany, NY, destroys 600 buildings.

1849 --
• California Gold Rush begins.
• San Francisco. A conflagration caused by arson results in $12 million in property damage. Following the fire, vigilante groups are organized to patrol the city and watch for more arsonists.
• May 17. A conflagration that began on the ship "White Cloud" destroys twenty-six river boats, 418 buildings, and kills twenty in St. Louis, Missouri

1850 --
• Fire destroys 400 buildings in Philadelphia and kills thirty-nine people.

1851 --
• May 4. The worst of six big incendiary fires that swept San Francisco between December, 1849, and June, 1851. This one destroys almost the entire city.

1852 --
• First fire-alarm telegraph central office and street box system inaugurated at Boston.
• Patent issued for first sprinkler-perforated pipe system. This was the first recognized installation of fire protection equipment.

1853 --
• Latta Brothers steam fire engine, "Uncle Joe Ross," revolutionizes firefighting. Cincinnati, OH becomes the first American city to replace volunteers with the horse-drawn steam fire engine, and to form a paid fire department.
• December 27. Great Republic, biggest clipper ship ever built, burns in New York on eve of her maiden voyage.

1854 --
• Jennings Building fire causes four deaths in New York City.

1856 --
• Aurora, IL forms its first fire company, "Young America Fire Engine Company No. 1" and a hand pumper and hose cart are delivered.

1857 --
• St. Louis, MO forms the second fully paid steam fire department in America.

1858 --
• Crystal Palace fire, in New York City.

1859 --
• Baltimore, MD established its first paid fire department.

1860 --
• January 10. Pemberton Mills Fire in Lawrence, MA.; 115 killed.
• February 2. Elm Street Tenement Fire in New York City; 200 killed. Laws requiring fire escapes were passed as a result of this fire.

1861 --
• Milwaukee, WI establishes a paid fire company.
• Washington, DC fire department becomes a fully paid organization and installs a fire alarm telegraph.
• Fort Sumter attacked. Civil War begins.
• New York City firefighters organize the first Fire Zouaves regiments and leave for the battlefront.
• December 11. Most of Charleston, SC is destroyed by fire.

1863 --
• Battle of Gettysburg.
• July. Draft riots in many United States cities.
• New York City draft riots occur.
• New York City becomes paid fire department.

1864 --
• November 8. "Southern Conspiracy" to burn New York City.

1865 --
• General Robert E. Lee surrenders.
• April 14. President Abraham Lincoln assassinated.
• April 27. S.S. Sultana explodes in Mississippi River; 1450 killed.

1866 --
Atlantic Cable successfully laid.
July 4. Great Fire in Portland, Maine. Firecracker starts blaze that destroys 1500 buildings.

1867 --
• Alaska purchased from Russia.

1869 --
• Railroad spans America, coast to coast.
• The City of Aurora reorganizes the fire department. A new steam fire engine is purchased for the East Side district. A high-pressure Holly Pump water system is installed and the "Holly Hose Company" is organized to protect the West Side.

1870 --
• "Boss" Tweed era of corruption in New York at its height.
• Philadelphia PA gets a paid fire department.
• Daniel Hayes, a San Francisco fireman, develops the first successful aerial ladder truck.

1871 --
• First volunteer firefighting unit organized in Los Angeles, CA.
• Rubber-lined, cotton-jacketed, fire hose begins to replace the riveted leather hose.
• October 8. The Great Chicago Fire, 18,000 buildings burned, over 200 lives lost. Help comes from eight states to battle the two-day conflagration. Aurora, IL sends a steam fire engine and crew of eighty-five firefighters.
• Also on October 8, forest fires destroyed the town of Peshtigo, WI (killing over 1,000) and raged through parts of Michigan.

1872 --
• November 9. Great Fire of Boston.
• Great Fire of Boston destroyed 776 buildings and one square mile of the business district.
• More than seventy insurance companies went bankrupt as a result of the Great Fire of Boston. The companies that survived formed the National Board of Underwriters and established safeguards for insurance companies to follow.

1873 --
• Quick-hitch collar and hames invented by Charles Berry a fireman in Cambridge, MA.
• Suspended harness invented by firemen in St. Joseph, Missouri.

1873 --
• First sliding poles, (made of wood), were installed in some New York engine houses.

1874 --
• Automatic sprinklers introduced.
• First high-pressure water system for fighting fires installed in Rochester, NY

1876 --
• Telephone invented.
• Big-league baseball is organized.
• December 5. Brooklyn Theater Fire; 295 killed.

1879 --
• Edison invents the incandescent lamp.

1880 --
• Frederick Grinnell improved upon the automatic sprinkler. Insurance companies cut rates to businesses installing Grinnell sprinklers.

1889 --
• Henry Ford builds his first car.
• May 31. The Johnstown Flood, 2200 people killed.
• June 6. Fire destroys thirty-one blocks in center of city and along the waterfront in Seattle, WA.

1894 --
• Moving-picture machine invented.
• New Central Fire Station (now the home of the Aurora Regional Fire Museum) is completed in Aurora, IL.

1895 --
February 14. Fire at Lynn, Mass., destroys 300 buildings in center of city, mostly factories.

1897 --
• Klondike Gold Rush.

1898 --
• Spanish-American War.

1900 --
• June 30. Fire sweeps through the Hoboken waterfront; 400 killed.
• September 8. Galveston, TX Flood kills 5,000.

1901 --
• May 3. Fire destroys 1700 buildings in Jacksonville, Florida.
• Marconi transmits the first wireless message across the Atlantic.
• President McKinley is assassinated and Theodore Roosevelt becomes President.

1902 --
• Wright Brothers make first airplane flight.
• February 8. Fire burns 525 buildings in Paterson, New Jersey.

1903 --
• December 30. Iroquois Theater fire, Chicago, IL (602 deaths and 250 injuries).

1904 --
• Successful breathing apparatus invented but not adopted for a number of years.
• February 7. Great Fire at Baltimore. Eighty downtown blocks, 1343 buildings burned. The Baltimore fire raised national attention for the need to standardize fire hose couplings and screw threads.
• June 15. General Slocum, an excursion steamer with combustibles on board, catches on fire while cruising in New York's East River. 1,030 lives were lost, mostly children. This fire lead to inspection of ships in New York Harbor.

1906 --
• April 18. San Francisco earthquake and fire destroys 28,000 buildings; Over 492 deaths and $350 million in property damage.

1907 --
• Gasoline-powered motors and pumps begin to appear in the fire service.
• Underwriters Laboratories initiated its factory inspection service and began to issue labels for "approved devices."
• Invention of first pumper with a single engine to do both driving and pumping.

1908 --
• March 4. Fire at Lakeview Grammar School in Collinwood, OH. 175 children and one teacher are killed.
• April 12. Conflagration in Chelsea, MA burns 3,500 buildings and kills eighteen.

1910 --
• Fire in the Nelson Morris Company Chicago, IL. Twenty-one firemen killed by falling wall.
• New York City fire, Grand Central terminal yard.
• Transport of hazardous materials is quickly becoming a great hazard.

1911 --
• Aurora, IL places its first motorized fire engine into service.
• March 25. Fire at the Triangle Shirt Waist factory in New York City. Over 150 killed, mostly young women. This fire incited the public and labor against sweat shops and child labor.
• New York City created Committee on Safety which led directly to Safety to Life Committee of National Fire Protection Association. (the NFPA).

1912 --
• Equitable Building fire in New York City.
• Titanic sinks, 1,513 lives are lost.

1913 --
• Fire destroys 518 buildings in Hot Springs, ARK
• Binghamton Clothing Factory fire results in new standard for building exits (35 deaths).
• Single-driving and pumping engine developed to efficiency.
• Life Safety Code® is established.

1914 --
• June 25. 1,600 buildings are destroyed when fire sweeps through Salem, MA.
• Panama Canal opened.
• World War I begins.

1915 --
• Telephone service from New York to San Francisco is initiated.
• The S.S. Lusitania is sunk.

1916 --
• March 21. Fire burns 1440 buildings, nearly the entire town of Paris, Texas.
• March 22. Fire in Nashville, TN burns 648 buildings, mostly dwellings.
• March 22. 682 buildings burn in downtown Augusta, GA
• July 30. "Black Tom Pier" fire and explosion in New Jersey.

1917 --
• Ammunition storage explosion kills 1,500 in Halifax, Nova Scotia.
• Eddystone Ammunition Corporation explosion, Eddystone, PA (133 deaths).
• April 6. United States enters World War I.
• May 21. Simultaneous fires destroy 1,938 buildings in Atlanta, GA

1918 --
• International Association of Firefighters Union organized.
• October forest fires in Minnesota sweep through fifteen townships and destroy and kill 559.
• November 11. Armistice is signed, World War I ends.

1919 --

• Two-platoon system begins in New York City. Many cities had already adopted shorter hours for firemen, others were to follow soon.

1920 --

• Prohibition in effect.
• September 16. Wall Street explosion; 40 killed.

1921 --

• New Stutz motorized fire engine is delivered to Aurora, IL (this engine is now on display at the Aurora Regional Fire Museum)
• California passed a law forbidding wooden shingles on roofs but pressure from the roofing industry brought a repeal.

1922 --

• President Harding issued first Fire Prevention Week proclamation.
• Standards for Municipal Fire Alarm systems are adopted.
• December 8. Fire destroys 30 blocks in the center of Astoria, Oregon during a rainstorm.
• December 20. Last horse-drawn engine in New York City is retired.

1923 --

• Cleveland School Fire in Beulah, SC, 77 deaths.
• Tokyo and Yokohama earthquake and conflagration, Japan (91,344 deaths).
• United States Chamber of Commerce sponsors National Fire Waste Council with contest for fire prevention in cities by local chamber of commerce.
• Fire prevention education in schools required by thirteen states.
• September 17. Brush fire roars through Berkeley, CA and

destroys 640 buildings.

1926 --
• Shakespeare Memorial Theater burns in Stratford-on-Avon.

1927 --
• First sound movies.
• Lindbergh flies from New York to Paris.
• Annual Fire Prevention Week inaugurated.

1928 --
• Fire in Fall River, MA burns 107 factories and several business blocks.

1929 --
• May 15. Cleveland Clinic Fire. 125 are killed when X-ray films burn and give off poisonous fumes. As a result of this fire, laws require the use of "safety film."
• Stock Market Crash. The great depression had devastating effects on fire department budgets resulting in losses of manpower and stations. Arson fires increased.

1930 --
• Ohio State Penitentiary fire, Columbus, OH (320 deaths).

1930-1933 --
• Comprehensive fire loss study estimates that most fire deaths involving the very old and the very young occur in ordinary dwellings.

1934 --
• S.S. Morro Castle caught fire off the New Jersey coast, 134 killed.

• Conflagration in Hakodate, Japan destroys half of the city and takes 2,018 lives.
• Woolworth building fire in Aurora, IL. Three firefighters are killed and six others were injured when a wall collapsed.
• May 19. Chicago Stockyards Fire.
• Fire destroys 20 blocks in center of Nome, AK on September 17.

1937 --
• Consolidated School fire in New London, TX gas explosion (294 deaths). This fire focused attention on the need to safeguard buildings not subject to municipal ordinances.
• German Zeppelin Hindenburg burned as a result of an engine spark igniting flammable hydrogen (36 deaths).

1939 --
• World War II begins. Incendiary bombs dropped on European cities.

1941 --
• December 7. Bombing of Pearl Harbor resulted in loss of 2,383 lives and 19 American ships. Other losses - $25 million damage to aircraft, $25 million damage to buildings, supplies and ammunitions, 88 ships damaged and 960 persons missing. Pearl Harbor experience led to fire training program of the US Navy.
• Japanese and European cities continue to be destroyed by war conflagrations.

1942 --
• November 28. Coconut Grove Night Club fire, Boston, MA (492 deaths). This fire showed the need for practical improvements. UL stepped up research on combustibles in public places.

1943 --
• "Rain of Terror," in Hamburg, Germany. Fire storms killed 60,000 to 100,000 people.

1944 --
• July 6. The Ringling Brothers and Barnum & Bailey circus tent catches fire in Hartford, Connecticut. 167 are killed and nearly 500 are injured.
• Gas explosion and fire at the East Ohio Gas Company in Cleveland, Ohio causes 130 deaths.
• Munitions Depot explosion in Port Chicago, CA, 300 deaths.

1945 --
• Dresden, Germany, explosive incendiary bomb raid 300,000 deaths.
• Eastern Air Lines DC-3 near Florence, SC, fire in flight (22 deaths). Empire State Building struck by aircraft (14 deaths).
• Germany surrenders. Japan surrenders after atom bombs are dropped.

1946 --
• June 5. Sixty-one die in a fire at the LaSalle Hotel fire, Chicago, IL.
• December 7. Winecoff Hotel fire in Atlanta, GA, 119 deaths.

1947 --
• The Morkoy, petroleum tanker fire in Los Angeles, CA. Fireboats used to fight miles of waterfront fire.
• April 16. Ammonium nitrate being loaded on the S.S. Grandcamp explodes in Texas City, TX. More than 600 are killed, including entire membership of volunteer fire department.
• Centralia Coal Co., Centralia, IL dust explosion (111 deaths).
• October 23. Forest fires burn in Maine burn 1,200 buildings and

kill 16.

1948 --
• DC-4 accident at Chicago Municipal Airport (12 deaths). This fire recognized the need for specialized aircraft rescue and fire fighting vehicles for airports.
• The US Postal Service issues a three-cent stamp honoring volunteer firemen

1951 --
• Sparky the Fire Dog created as a symbol of fire prevention for children.

1954 --
• Cleveland Hill School fire, Cheektowaga. NY, 15 deaths.
• Oil refinery fire in Whiting, Indiana resulted in $16 million loss, explosion and boil over.
• Junior Fire Department in Los Angeles, CA organized.

1958 --
• Our Lady of Angels School, Chicago. IL, rubbish fire spread through open stairway (95 deaths).

1965 --
• Multiple fires resulting from Watts riots in Los Angeles, CA. After 34 lives were lost, special gear was used to protect fire fighters as a defense.

1966 --
• Chief E. J. Bauman establishes the Aurora Fire Department Museum. Lieutenant Charles 0. Goodwin is appointed as Curator.

1967 --
• McCormick Place fire, Chicago, IL
• Apollo Space Capsule fire, Cape Kennedy, FL, 3 astronauts died.
• A fire resistive fabric, Nomex,® came into existence.
• Forrestal aircraft carrier fire off coast of Vietnam, 131 deaths.

1974 --
• Aurora, IL firefighters begin Emergency Medical Technician (EMT) training at Mercy Center Hospital.

1975 --
• A 727 plane crash and fire in New York. Spilled fuel ignited after crash, 113 deaths.

1976 --
• Social Club fire, Bronx, NY kills 25
• Nursing home fire in Chicago, IL kills 24.

1977 --
• Fire at the "Beverly Hills Supper Club" in Southgate, KY results in 165 deaths.

1980 --
• MGM Hotel fire, Las Vegas, NV 85 deaths.
• Stouffer's Inn hotel fire, Harrison, NY, 26 deaths.

1981 --
• Hyatt-Regency Hotel walkway falls, killing 114 in Kansas City, MO

1995 --
• Oklahoma City Bombing, 169 Dead.

2001 --

• September 11. Terrorist attack the World Trade Center in New York City and the Pentagon in Washington DC. 189 are killed in the Pentagon attack. In New York, the 110 story World Trade Towers collapse killing nearly 3,000 including 344 firefighters and 87 police officers. A third hijacked plane crashed in rural Pennsylvania.

Bibliography

Brannigan, Francis L. SPFE and Carter, Harry R., Ph.D. The American Fire Service - Fire Disasters: What ...September 1998, < www.firehouse.com/magazine/american/disasters.html > (20 September 2005).

Buff, Sheila. Fire Engines in North America 1991. 1st edition. Secaucus, NJ
The Wellfleet Press. 1991 (20 December 2005)

Calderone, John A. The American Fire Service -Fire Apparatus Past and Present September 1998
<www.firehouse.com/magazine/american/apparatus.html> (23 September 2005)

Calfee, Mica. Fire Department History, Terminology and Tactics, Fire Department Rank Insignia. 2004
<www.riotacts.com/fire/reporters.html> (20 November 2005)

Ditzel, Paul C. and Smith, Dennis. History of the Leather Helmet 1976-1978 <www.firefightersrealstories.com/tidbits.html> (6 January 2006)

Hashagen, Paul. The Development of Fire Hose 1997 -2005 <www.firefightercentral.com/history/development_of_fire_hose.htm >
(10 December 2005)

Hashagen, Paul. The Development of Breathing Apparatus September 1998,
<www.ksffa.com/development_of_breathing_apparat.htm> (2 October 2005)

Hashagen, Paul. Firefighting in Colonial America. September 1998 <www.firehouse.com/magazine/american/colonial.html> (15 December 2005)

Lewis, David. History of the Fire Service. January 2003
<www.auroraregionalfiremuseum.org/history/afdhistory.htm> (28
August 2005

Page, James O. The Paramedics 1979
<www.info.jems.com/paramedics/ch12a.html> (2 January 2006)

Zatz, Arline. History of the Star of Life, The Emergency Medical
Care Symbol. January 2002
<www.hultgren.org/library/misc/starlife.htm> (12 September 2005)

Unknown. A Brief History of the Hydrant January 2003
<www.firehydrant.org/pictures/hydrant_history.html> (20 October
2005)

LaVergne, TN USA
17 December 2009
167350LV00005B/34/A